Incredibly American
Releasing the Heart of Quality

INCREDIBLY AMERICAN

RELEASING THE HEART OF QUALITY

MARILYN R. ZUCKERMAN
AND
LEWIS J. HATALA

ASQC Quality Press
Milwaukee, Wisconsin

Library of Congress Cataloging-in Publication Data

Zuckerman, Marilyn R.
Incredibly American: releasing the heart of quality / Marilyn R.
Zuckerman and Lewis J. Hatala
     p. cm.
   Includes bibliographical references.
   ISBN 0-87389-192-9
   1. American Telephone and Telegraph Company—Management.
2. Total quality management—United States.   3. Corporate culture—
United States.   I. Hatala, Lewis J.   II. Title.
HE8846.A55Z83   1992   92-17210
384'.06'573—dc20   CIP

109876543

ISBN 0-87389-192-9

Acquisitions Editor: Jeanine L. Lau
Production Editor: Mary Beth Nilles
Marketing Administrator: Susan Westergard
Set in Galliard and Optima by Montgomery Media, Inc.
Cover design by Montgomery Media, Inc.
Cover photograph by Scott Axtman
Printed and bound by BookCrafters

For a free copy of the ASQC Quality Press Publications Catalog, including
ASQC membership information, call 800-248-1946.

Printed in the United States of America

Printed on acid-free recycled paper

ASQC
Quality Press
611 East Wisconsin Avenue
Milwaukee, WI 53202

*This book is affectionately
dedicated to the people we love.*

*They have unselfishly and enthusiastically supported
us and remain our greatest source of inspiration.*

CONTENTS

FOREWORD

Quality doesn't mean the same thing to Americans as it does to the Japanese, Germans, French, or members of any other culture. What motivates a Japanese worker won't "turn on" an American. The Germans have an obsession with standards that is missing for Americans. The French have a luxury component in their quality definition that Americans don't recognize.

Achieving perfection and the pursuit of zero defects is "un-American" in a very real sense. However, going where no man or woman has gone, reaching for and achieving the impossible, are definitely American. Not only was the United States founded by people seeking to achieve the impossible, but America continues to be the magnet for people all over the world who want a new start, a place to achieve their impossible dream.

This book is about *American* quality—what it is, what it isn't. It concentrates on one dimension of quality that has long been overlooked: emotions. The emotions associated with quality in America are unique to our culture. The diversity of world cultures offers us many creative solutions and wonderful opportunities for learning, but the best culture to learn from is our own.

When American business finally comes to grips with the emotional elements of American quality and translates that understanding into American action, watch out! History is filled with the consequences of Americans unleashing the power that comes more from our emotional energy than our engines.

In 1986 AT&T began a search for the uniquely American dimensions of quality. The discoveries that the authors and I and others at AT&T made during our study are recounted, in part, in this book. The process of discovering these truths is imbedded in biology, psychology, learning theories, and cultural anthropology. It comes from a synthesis of Konrad Lorenz's a work on imprinting, Piaget's work on the child's stages of maturation, Jung's approach to the strong notion of collective unconscious, and my research and writing on discovering and breaking the codes of cultural archetypes.

Many researchers and authors have explored these dimensions in an intuitive way. I am indebted to people like Ruth Benedict and Edward T. Hall for their work on Japanese culture, and to Levi Strauss for his emphasis on structure. They, among others, are the architects of the process that I use today. Their research and analysis have paved the way to understanding culture as a series of "mental highways" (neuronic pathways) imprinted at an early age through the function of our nervous system. These mental highways represent the cultural cognitive structure that is available to an individual in a given culture at a certain time.

As we know from the work of the French biologist Laborit, emotion is the energy that creates the imprinting. Each culture has its own "logic of emotion." When you push the right button, you fire up a mental neuronic circuit and generate (or reactivate) the emotional energy that was used to imprint the circuit. So when you know the code, you can turn on, or turn off, a culture.

Over the past 18 years I have developed a methodology for finding these structures through an archetype discovery process. Using archetyposcopes, we are able to discover the cultural cognitive structure (mental highways) available in a given culture. Archetyposcopes are analogous to microscopes. They don't create or invent anything, they reveal things, just as microscopes reveal microorganisms. At AT&T we applied this process to quality.

What makes this book so special is that it represents the only detailed reporting of our discoveries. This book's most important characteristic is that it is written not only from a professional point of view but from the authors' personal perspectives: what they saw, heard, and felt. Their emotional experiences inform the reporting.

In many respects the authors' perspectives will be most useful to the general reader. Others more familiar with the works already cited and with the German gestalt approach to the notion of structure or the new systematic thinking current in psychology may take exception to the personal narrative and wish for more objectivity. Let me assure these readers that the quality archetype has nonetheless been independently validated. Although not referred to in this book, the qualitative and quantitative work currently being done by the American Quality Foundation supports the findings of the quality archetype. In fact, using a different methodology with an independent researcher, the American Quality Foundation has duplicated many of the archetype findings reported in this book.

To the specialists or more skeptical readers in our midst: I trust that you will see this book as a timely reminder of the work that needs to be done to get America's quality journey back on track with strategies that focus on our cultural strengths.

When the discoveries of the quality archetype were first made in 1986, the "soft" areas of quality were being ignored by business. The pressure for systems, technology, tools, and other "hard stuff" was very strong. Only recently have companies, including AT&T, started focusing on the human side of quality. As this process unfolds, the quality story told by Marilyn Zuckerman and Lew Hatala deserves everyone's attention.

The emotional aspect of quality does not contradict any of the previous work done by the quality gurus. In a very Jungian way, this is not an either/or situation. It's more like opening a door to an inner world of hidden forces and treasures that, when fully understood, will magnify the work others have done before.

What this book offers is a deeper understanding of the underlying structures of how we think about quality as a culture and why we behave the way we do toward quality. For some, these discoveries will require a fundamental shift from where they are today. The findings may be discarded because they appear contradictory, incomplete, or foreign in a figurative sense. Or they may be put aside because there is no application immediately apparent. For others, the archetype findings will confirm much of what they already knew; the findings will remind them of why and how they think and behave the way they do. And for still others, this study will give a new natural and logical system or framework for thinking about and understanding quality.

We now have an opportunity to rediscover America and the American logic of emotion that underlies and supports our unique notion of quality. It is once again time to be proud to be an American, not for what we have accomplished but for what we are going to accomplish in our own special way.

Dr. G. C. Rapaille
Chairman, Archetype Studies International

ACKNOWLEDGMENTS

This book began with a shared dream. Before it was realized, we had lived through many false starts, several crises, and a few rock-bottom failures. But all along, our personal rewards came from reliving many experiences we had shared with family, friends, and colleagues, many of whom you'll meet within these pages. We hope your encounters with these people will enrich your lives as they have enriched ours. Without question, the dream of writing this book would never have become a reality without the support and encouragement of so many.

Many wonderful people at AT&T and other companies made important contributions. These are the people from whom we learned—and continue to learn—about the Olympics of the heart and about the important role that caring for one another plays in the quality of our lives at home and at work.

A number of our friends were kind enough to read our earliest attempts at a manuscript and to provide us with constructive comments and suggestions. These friends include Hub Evens, Doug Fernandez, Lisa Green, Kitty Holland, Linda Merritt, James Porter, Don Reser, Jim Riordan, Madeleine Risser, and Sherl Smith. Their help and their contributions have been invaluable.

We are especially indebted to Peppi Elona, Mary Sullivan, and Steve Wise, who were with us from our first prototype to our last draft. Their insights challenged us to reach beyond our earliest expectations.

We want to thank two special AT&T colleagues, Bob Kerwin and Larry Lewallen, and our many special friends at NYTV, who gave support, encouragement, wise counsel and friendly advice.

Because of their courage and pioneering spirit, there will always be a special place in our hearts for Lee McClary and Ray Peterson. In very different ways, they each took risks on our behalf, without which none of this work would have gone forward. They were willing to say yes and go against the tide of conventional thinking. When others were questioning the value of the study presented in this book and asking why, they were willing to say why not!

We are indebted to Dr. G. Clotaire Rapaille, who has added a new dimension to our lives and the lives of many others. His theory and research methods gave us a new set of tools with which to discover the American archetype of quality. He enabled us to make a major breakthrough in our understanding of the way Americans function that brought to light the critical elements previously missing from the human side of quality.

We want to thank Bill Idol for his many contributions. He supported the workshops in ways that helped transform the study into powerful learnings.

A word of acknowledgment delightfully goes to Jeanine Lau at Quality Press. She supported our early incoherent work, prodded us when we needed it, and provided pats on the back to keep us going. Thanks, Jeanine, for your faith in us.

We were pleased and honored that the American Quality Foundation (AQF) chose this work as an integral part of their current project, the Personal Quality Improvement Process.sm We thank Josh Hammond, AQF President, for supporting our work and offering us a vehicle for sharing it with many other American companies.

We offer special thanks and appreciation to our colleagues and friends with whom we have served on the AQF Steering Committee. We especially thank Joe Bransky, from General Motors, Denny Crouch, from the Federal Quality Institute, Mary LoSardo, from Met Life, and Joanne Imlay and Jim Gerard from NYNEX for their comments and suggestions for improvement. Their strong endorsement of this work carried us through a very difficult time.

We are grateful to Rita Lloyd and Leahrae Belk, both invaluable supporters. Facing an ongoing barrage of notes and scribbles, they performed the miracle of converting them into a typed manuscript.

Phil Scanlan, Quality Vice President at AT&T, spent long hours during his vacation reading and analyzing our manuscript. His comments and suggestions were penetrating and insightful and have served to improve the book immeasurably. During the many crises we faced Phil was always there giving unselfishly of himself as both our coach and mentor. For his help, his support, and his friendship we are deeply grateful.

We must pay special tribute to Don Riemer. A gifted writer and editor, Don transformed our manuscript into a far more readable and effective work. He could not have given more, even had this book been his own. His insight, clear thinking, and lucid writing are unsurpassed, and we remain forever grateful for his contribution. To Don—our heartfelt thanks.

And we want to thank each other for the mutual inspiration, encouragement, and support to "keep on keeping on" during the worst of times and the best of times. We certainly couldn't have done it without each other.

My biggest personal thank you goes to my children, Sarah and her husband, Brian, and Lisa and her fiancé, Bob. The time and effort you contributed to make this book the best it could be has touched me deeply. Whether it was on the sidelines coaching and cheering or in the very thick of it, reading, probing, thinking, and writing, you were always there for me. Once again, you have given from your heart and have made me very proud to be your loving mother.

—Marilyn

Special thanks go to my in-laws, Audrey and Harold DeSombre, for their continued confidence in me, and to my family. Kim, Tom, and Jeff, it was your willingness to review the manuscript countless times, offer loving support through thick and thin, and provide warmth, encouragement, and inspiration that truly produced this work. And I want to share my byline with my wife, Barbara. She worked with me from the start, editing, critiquing, and adding ideas and insights from her own experience. Thank you also, Barbara, for your unending patience, support, and love during all those hours I spent working on this book. You made it important to you because you knew it was important to me.

I love you all.

—Lew

*I had learned many English words and could recite part
of the Ten Commandments. I knew how to sleep on a
bed, pray to Jesus, comb my hair, eat with a knife and
fork, and use a toilet ... I had also learned that a person
thinks with his head instead of his heart.*

—Sun Chief

*We don't receive wisdom; we must discover it for ourselves
after a journey that no one can take for us or spare us.*

—Marcel Proust

You can close your eyes to reality but not to memories.

—Stanislaw J. Lem

INTRODUCTION

STUMBLING ON QUALITY

T his book grew from an unexpected collaboration between two people who originally inhabited widely separated places in the vast corporate labyrinth that is AT&T and who ordinarily never would have met. But divestiture and the growing pressures on the telecommunications business led to a series of events that brought our worlds together.

In considering the emotional ingredients that sustained the project that eventually gave rise to this book, we recall dominant flavors like frustration, disappointment, anxiety, and bleeding-from-the-eyes anger. But there was also thrilling discovery, warm respect, eager anticipation, and some wonderful floods of joy and hope. Our collaboration began during one of the hopeful periods.

MARILYN AVOIDS A MIPP

In late 1985, I thought I had perfected the art and science of finding a job within AT&T. To survive in those days, one had to keep away from the dreaded MIPPs, which haunted the corporate hallways like sly, administrative hit men; if they caught you it was all over. MIPP meant Management Income Protection Plan, which was a nice way of saying early retirement. It was basically a layoff with a very generous cash bonus, which gave the company a kinder and gentler way to lighten ship—and lighten ship they did. This period was close on the heels of divestiture, and reorganization was spreading through a shocked and panicky AT&T like scarlet fever. Entire divisions, directorates, and business groups were being consolidated, downsized, shuffled, or eliminated altogether. Job security had become a thing of the past.

In the course of 1985, I had four different assignments, all in the same corner of New Jersey, and learned one day that my current job was folding along with my director's entire organization. MIPP activity was time-dependent, which meant I had a limited window of time in which to find another position at AT&T before joining the ranks of the "Mipped."

I found a few jobs open, but none was appropriate. My nerves were frayed; I was sleeping poorly and lived on an unpleasant roller coaster of frantic high energy and dark depression.

One might ask: why all the fuss? Why not MIPP out, lie on the beach for a few weeks, and then look for a job someplace else, as so many others had? This is a tough one to answer. It has a lot to do with the way some of us had come to feel about AT&T.

Part of it is loyalty—the sort of feeling you might have about any organization with a long history, noble beginnings, and an inspired vision. I can imagine feeling this way about MIT, the New York Philharmonic, or the National Geographic Society. AT&T was the first to demonstrate the commercial and social power of high technology. AT&T brought the world transistors, undersea cables, and communications satellites. They changed the character of American life—it's hard to walk away from that sort of tradition.

Part of it is pride in having a chance to carry on that tradition. The feeling of being lost in a labyrinth is tempered by an awareness that you're a part of something unique, something that still has tremendous potential and before it's through, may change American life again, and maybe again after that.

Part of it is curiosity. Just when you think you've seen it all, AT&T demonstrates that it still has the power to amaze—one reason I stick around is just to see what will happen next, and perhaps to be a part of it. And part of it is pure, bullheaded stubbornness—an unwillingness to abandon ship even when it seems like a healthy idea. And perhaps most of all, part of it is hope—the hope for . . . something. AT&T is huge; it's impossible to see the whole company, and the confusion after divestiture added instability to an already incomprehensible structure. In 1985, AT&T was a wounded giant, but it was still alive and trying desperately to survive in a changing world. Back then, I couldn't tell whether it was bleeding to death or merely stumbling on the road to recovery, but I hoped it was the latter. I wasn't about to leave.

So, I kept looking for a new position ... and couldn't find one. About when my inner resources were at their lowest, I noticed an envelope that had appeared mysteriously on my desk (the angel who'd sent it remained a mystery for quite some time). The contents described a job I had apparently missed, one that was very appropriate. I applied just hours before the deadline, and a short time later became something called a Quality Manager. What I didn't know then (but know now) was that I had just lived through a culturally mandated American experience: crisis.

Installed in my new office as Quality Manager, AT&T Network Systems, I felt purged, hopeful, and ready to work. I learned that the position had actually grown from an apparent postdivestiture need: many Network Systems organizations had expressed a desire for guidance on business quality. Quality methods were already well established at Bell Labs and AT&T's manufacturing sites, and other organizations were now eager to learn how to "do quality." Several task forces, those versatile knights on AT&T's chessboard, made recommendations about how to proceed. Finally, the Executive Quality Council approved the creation of the Quality Manager position, reporting to Ray Peterson, Quality Director.

The folks in Network Systems were ready and eager for quality, and my job had been specifically created to bring it to them. Remarkable, I thought. What more could a manager want than to develop a comprehensive quality education program for more than 100,000 people? I was thrilled! I dove into my work with the joy of the recently rescued and the enthusiasm of the newly converted.

I was charged with implementing quality training from the top down, so I worked with Ray and another task team to design a

Quality Leadership Conference (QLC) for roughly the top 150 managers in Network Systems. The QLC would give our leadership a fresh, sharp focus on quality. It was clear that quality had become a critical parameter in our customers' minds; in that arena, AT&T wasn't always making the grade. It was hoped that the QLC would enable the Network Systems leadership team to look ahead to where AT&T and its competition might be in five or ten years. Tough goals were to be set to create a sense of urgency about quality that would cascade down through the organization. This was an important step. The QLC would send the first big message, kick off our long-term initiatives, and shape all Network Systems quality activity for the coming years.

The QLC became my life, and as time passed it also became a real source of frustration. The importance and urgency I attached to quality issues were clearly not felt by all those around me, and slowly my enthusiasm turned to disappointment. I worked for weeks building, refining, and molding the Conference into something powerful and compelling. After much talking, listening, thinking, and brain-picking, I began to throw a series of conference designs at my boss. Finally, one night past midnight he caught one, looked it over, and said, "That looks good."

I felt great—no, more than great. I felt elated, on top of the world, and above all I wanted to keep going. My weary frustration vanished and I felt a huge resurgence of energy that I can only describe as amazing; it dumbfounded me.

And by the way, what was my boss doing in the office after midnight? For a moment, the waiting area outside a hospital delivery room flashed through my mind. Could he be waiting for the birth of the QLC? Well, I had labored and finally delivered, and he seemed to like the baby. We never took the time to celebrate—actually, it never occurred to either of us that there was cause for celebration. We could only look ahead to the next challenge, but at least we felt ready.

With considerable pride and confidence, I unwrapped the baby in front of the vice presidents at the next Executive Quality Council meeting. I looked for a response, waiting for them to love the child as I had grown to love her or him, whatever. Well, the scene as I remember it was grim—the baby did not please.

Some never bothered to look, some were asleep, some said "She's too big," others, "She's too small," "She's not healthy," or, and this one dominated, "We're not ready for her." Not ready for quality—that was too much. At least I hadn't bought cigars.

We became bogged down over details. Finally, the Council chairman asked for a show of hands: "How many approve?" Most raised their hands on cue but with visible reluctance, as if they were voting for an incinerator in their hometowns. My boss and I left, somewhat deflated, and with the impression that we carried a bruised and wounded baby. Nonetheless, we were expected to go on.

In conjunction with the QLC, I was involved in a host of varied other quality activities and was trying to implement a number of quality improvement practices, throughout the organization; they weren't working well either. This was as confusing as it was frustrating. Methods, practices, and systems were selected that are the lifeblood of Japanese business, things that were proven in the country that was beating one American company after another. Why weren't they working? Despite the organization's early enthusiasm for quality, everything attempted met with apathy, disinterest, or benign neglect. Even the groups that seemed to be genuinely trying to cooperate didn't do well with the new quality tools. Hondas and Toyotas began to take on mystical significance.

During this low point, I had a chance to talk with Dr. G. Clotaire Rapaille, a cultural anthropologist and psychoanalyst turned archetypologist™, with whom I had worked some years before. Knowing his ability to answer the seemingly unanswerable by exploring the details of different cultures, I spoke with him about my dilemma. He believed his study methods were likely to uncover some of the missing pieces in my approach and would explain why we were having trouble motivating people to "do quality."

We decided to take a shot at applying Dr. Rapaille's methods in Network Systems. With great trepidation and little confidence, another baby was conceived, this time in quiet secrecy—just a few close, understanding relatives were told about the project. These were some people on the QLC project team who were the most committed and the most disturbed by the apparent lack of support. With their encouragement, the boss agreed that the baby looked pretty good. "Okay, try to grow the baby," he said. With his support, we all felt a resurgence of energy—not as strong as the first but enough to energize us for the new project.

Carefully and cautiously, we developed a new proposal and project plan. The scope of this plan was really quite narrow. We just wanted to use Dr. Rapaille's approach to help us get past the roadblocks that were threatening the QLC project. Initially, we saw it as a brief divergence to explore our American culture, followed by a return to our QLC tasks with the missing pieces in hand. In a short time, the proposal was complete, lacking only the boss' signature on Dr. Rapaille's contract. When the time came, he refused to sign it and offered no explanation.

The team members who had worked on the proposal were livid and disbelieving; the situation was almost surreal. The boss knew all the details, all the background. He had attended most of the planning meetings and had been among the most fascinated by Dr. Rapaille's approach. All along, he had been an ally, but suddenly he spun about-face.

Our response was well out of proportion to his calm refusal. All the pent up frustration, all the absorbed defeats, all the weekends, late nights, and bad coffee of the past few weeks welled up and exploded in an emotional counterattack. The battle was brief—he surrendered and signed the contract, with a somewhat indifferent expression on his face.

In retrospect, I believe his refusal was understandable. Was he willing to take such a big risk? Was he willing to support a relatively expensive project, one like none other he'd been a part of during his 20 years with the company? If it failed, would he suffer the consequences? And if it succeeded, would anyone care?

The cultural study proceeded separately from the QLC. It moved along quietly in the background, with no fanfare and no publicity, something like a skunk works project. We didn't realize it at the time, but feeling and acting like a skunk works group gave all of us a lot of energy. The faint air of secrecy and urgency fueled our passion to succeed.

At no other time and on no other assignment had Ray or I seen a team so totally committed. We often worked until close to dawn, analyzing the study's findings. Picking through the pieces was like panning for gold, identifying, prodding, and arranging the precious flakes and nuggets as they appeared one by one. The findings alone were surprising, delightful, and fascinating. What we didn't know was that an even bigger shock was waiting for us and would soon arise from a completely unexpected quarter.

LEW SEES A PATTERN

Early in 1985 I had never heard of Marilyn Zuckerman or Dr. G. Clotaire Rapaille or the QLC. I worked in a very different part of the Network Systems organization, a side of the business that people at AT&T's corporate nucleus rarely see: manufacturing. While Network Systems' administrative heart lay nestled in the Jersey hills, the Copper Shop lay smack in the Heart of Dixie, in the industrial rim 10 miles northeast of Atlanta.

Copper meant copper telephone cable, and before divestiture AT&T had used a lot of it. Located at the Western Electric Atlanta Works manufacturing unit, the Copper Shop produced cable for a variety of telephone and data communications needs and now makes similar cable for fiber optic applications. I was the Shop's operations and engineering manager, concerned with product and process engineering, production schedules, material deliveries, union negotiations, and a constant barrage of minor, day-to-day details. It often reminded me of running a small country, maybe Liechtenstein.

In January 1986, I was transferred out of my operations slot and into the position of Quality Control Manager for the entire Atlanta Works, which involved about 3,500 people. To be frank, I wasn't happy about it. While there were new challenges for me in this position, I felt the move was premature—the Copper Shop had faced some tough problems in the previous few years and was just beginning to turn around. I had attended a Deming management seminar the previous March and so far hadn't had much of a chance to apply the fascinating tools this experience provided. I felt Dr. Deming's recommendations to redefine business organizations, improve productivity, and modify virtually all accepted concepts of work were just what the Copper Shop needed. I had begun to see my job and my workplace very differently and wanted to continue putting these new tools to work. Instead, I was being pulled away to a new assignment. I was afraid this change would retard the Shop's momentum, something we couldn't afford.

To add to the pressure, I was also tapped to represent my line of business on a project team for the then rather mysterious QLC. I knew of it only as a new quality program being developed in New Jersey by someone named Marilyn Zuckerman. It turned out to be the first major event in a long-term quality plan for Network Systems.

Now don't get me wrong. While I applauded the people who did this kind of work, it was definitely not my cup of tea. My specialty was engineering management; I liked it and I was good at it. But there seemed to be no choice, and who knows, maybe these folks were onto something. I promised to give it my best shot, despite the need for many plane trips between Newark and Atlanta. With the new project team, I began to prepare my piece of the QLC.

One of the Conference segments called for an in-depth analysis of quality success stories at AT&T. The intent was to explore a number of these stories to learn the pattern of success and to discover the details and conditions that lay behind it. When the call went out for success stories, Marilyn received more than 200 detailed written descriptions. My people sent in a story about the Copper Shop's recent turnaround.

Before our full plan for the QLC was implemented, the entire project ran into a long, tedious, and frustrating series of snags. When we needed organizational support beyond the project team, we couldn't get it. The decision makers who set the direction for the QLC were unwilling to move on it and weren't supporting the goals they'd set. The team members continued to work under these constraints, believing we were empowered to bring about change even though we found little of the needed help within our organizations. Support on all sides continued to dwindle.

In July, without any warning, much of the project came to a screeching halt; the Quality Council had indefinitely postponed the QLC. Many of the team members had been devoting full time to this project and were devastated at this turn of events. No one offered any explanation, and that's when the frustration, disappointment, anxiety, and anger set in. But although the QLC was in serious jeopardy, the success story module was still alive.

It was a uniquely painful experience to see the program failing. During the preceding few months, I had come to believe in its merits and was finding the process more stimulating and personally rewarding than I had expected. I was especially intrigued by the success story probe and anxious to break the code of quality success. But now it looked like the whole program might never get off the ground.

We kept working, though my enthusiasm dropped noticeably. When I picked up the phone one day to learn that the Copper Shop story was the one selected for the Conference, some primal brain centers took over and I did in fact jump for joy. The Copper Shop story contained elements of both quality improvement and business turnaround—I knew the elements of the story were real and powerful but didn't actually expect it to stand out in a field of 200.

I was told that of all submissions, ours was the only one about people. Other stories featured process and product data to substantiate quality improvement. Our story featured our people and gave particular attention to their sincere appreciation for their managers. None of the other stories had grabbed the selection team as ours had.

What apparently did the trick was the following statement, carried in the submittal letter that accompanied the story: "The production specialists held a management appreciation breakfast for their supervisors and managers, to thank them for changing." I later learned that the selection team kept coming back to our story during the review process because of this one statement.

This was outstanding. I immediately passed the word to the people at the Copper Shop. Telling them was like finding out myself all over again—more jubilation. There aren't too many things in the corporate world that convey a deep feeling of honor, but this particular distinction did, no question. We were thrilled and couldn't wait for the Conference, regardless of the form it would finally take. I continued preparing for the QLC, now with the added excitement and considerable fascination of exploring every detail of the Copper Shop story. We relived the whole episode and virtually dissected the pattern of events piece by piece, emotion by emotion. Gradually, like paleontologists chipping sandstone from a priceless skeleton, we uncovered the structure at the story's heart.

When the remarkably simple pattern, with its strong emotional dimension, became clear, I experienced a moment of pure, intense fascination, almost of intrigue. But there was an even greater marvel waiting for me—a surprise that went off like a bomb and changed my life.

I had known for a while that, in addition to the Conference and all its ancillary activities, Marilyn was working on some sort of cultural study, something to do with American quality. I didn't know much about this side of her work; it had the lowest of profiles, like some sort of covert CIA plot. Well, I found out soon enough. In a demonstration of unplanned synchronism that still makes me grin, she produced the final results from her cultural study within a few days of getting the report on the pattern of events in the Copper Shop story.

They were the same.

A key and its clay impression, a building and its shadow, a tree and its reflection in a still lake could not have matched more closely. The cultural study and the Copper Shop investigation had been performed by different groups, had looked at different phenomena, with different intentions and methods, but had uncovered virtually identical patterns of human emotion, response, and experience. Marilyn and I quickly realized our work was just beginning. The next few years found us becoming close friends, spiritual warriors, and partners in discovery; exploring the misty recollections of early childhood and the ponderous trends of American business, disaster, and victory; and ultimately collaborating on this book.

Nobody can do for little children what grandparents do. Grandparents sort of sprinkle stardust over the lives of little children.

—Alex Haley

— ✧ —

Life is not a "brief candle." It is a splendid torch that I want to make burn as brightly as possible before handing it on to future generations.

—Bernard Shaw

— ✧ —

Life is not dated merely by years. Events are sometimes the best calendars.

—Benjamin Disraeli

PROLOGUE

SCRATCHING AT THE FOSSIL RECORD

The patterns that emerged from the cultural study and the Copper Shop turnaround tell a familiar story. The story could be about Kathy, a budding commercial artist from San Francisco, who has just begun storming the downtown agencies with a flashy portfolio and a BA from Pratt. It could be about Carlos, a rancher born near El Paso in 1872, who fathered nine children and died crossing the Rio Grande at the age of 50. Or it could be about Franklin, a high school dropout who eventually became a lawyer at 35, and at 58 will be elected mayor of Morgantown, West Virginia.

This story could be about any man or woman, from any city or town, and any generation in the last 200 years. None of the background matters, except for two details, and they matter very much indeed. The first is that the hero or heroine of the story is about six years old when the story begins. The second is that he or she is an American.

It's 1955 and Jeff (our hero) is in the first grade. He's a good kid: plays hard, takes care of his dog, doesn't get in too many fights, and loves his parents like crazy. Jeff always tries to please, to do his best.

At school one day, Jeff and his classmates received a homework assignment. To Jeff, it seemed like a fairly straightforward piece of work, so when he got home he didn't start it right away. He played for a while, took his dog for a walk, and by the time he sat down to tackle the assignment, he had trouble remembering exactly what the assignment was. The teacher's instructions had seemed clear and simple in the classroom, but now, try as he would, he couldn't quite remember what he was supposed to do. The rules were no longer clear. But Jeff always tried to do his best, so he plunged forward and produced what he thought his teacher wanted.

When he had finished his homework, Jeff thought it was beautiful, really beautiful. He felt the same emotions that would later be stirred by the Grand Tetons at sunrise, Charlie Parker's saxophone solos, and his wife's profile. His homework also filled him with pride. This was his beauty—he had made it. People would be impressed with it and with him.

Confidently as can be, Jeff went to school the next day and handed in his homework. As the teacher collected everyone's assignments, he noticed that his was very different from what all the other kids had done. At first this made him feel great, and he figured it must mean that his was just that much better than theirs. But as he noticed the uniformity of the other kids' stuff, a scary feeling began to grow inside him—he wondered if maybe he had done the assignment wrong and everyone else had done it right.

The next day, his teacher passed out the graded homework assignments, and Jeff realized he hadn't done it right at all—he'd made a lot of mistakes. What he had thought was beautiful was just all wrong. To make matters worse, the teacher held up his paper to show the whole class everything he had done wrong. Jeff felt sick and a little dizzy. He wanted to hide, to escape all the eyes he knew were staring at him. His face was flushed and hot, his chest felt tight, and he was terrified that he might cry. His teacher gave the homework back to him and said he'd have to do it over.

He went home that afternoon feeling physically and emotionally awful. He didn't talk to anyone. He ran past his Mom to his favorite safe place. He sat in the dark, hiding, occasionally switching on a flashlight to page through a comic book. But mostly he just sat in the dark.

Later that day, Jeff's grandfather came to visit and found him in his hiding place. He peeked in politely and asked Jeff if he wanted to go for a walk. Jeff found this question very hard to answer. Part of him wanted to just stay in the dark and hide, maybe forever, but another part of him wanted very much to come out and tell his grandfather what happened at school. He took his grandfather's hand. They went outside and started walking down the street.

Jeff told his grandfather what happened with his homework. His grandfather never interrupted. He just listened, and every now and then he asked a simple question. Jeff became even more upset while telling the story and was crying by the end of it, but he never let go of his grandfather's hand.

After he was finished, they walked along silently for a minute, and then Jeff's grandfather started telling him a few stories of his own. He told Jeff stories about when he had been a kid himself, and some other stories about when Jeff's father had been a kid. They were simple stories, mostly about getting something wrong, feeling awful, and then fixing it; there always seemed to be a second chance. Soon, Jeff was feeling a lot better. His grandfather suggested that he give the homework another try. Suddenly, and very much to his surprise, Jeff felt an unexpected yearning inside. He really wanted to try that homework again—in fact, at that moment he wanted to try it again more than anything in the world. He practically dragged his grandfather back to the house.

Jeff did the homework again, and although he couldn't explain it, the emotion that filled him as he worked was the taste of impending and inevitable victory, the thrill of the tradesman, the teacher, the scientist, or the manager joyfully attacking a problem that he finally understands.

This time Jeff was successful. This time his teacher held up his homework for the whole class to see because it was the best, because he had done it exactly right, and to Jeff it was more beautiful than ever. When he left school, he felt 10 feet tall. He told his mom the whole story, repeating many times the moment when his teacher held up his homework. She was so proud that she made his favorite dinner and dessert, and that night the whole family celebrated his success.

Just before he went to sleep that night, Jeff went back to his favorite place. It didn't seem like a hiding place now. Instead, it made him think of great things in his future. He didn't know exactly what they were, he just knew that someday he would do great things, great and beautiful things. And maybe from now on he'd be able to do them right the first time. He sure would try.

— ✧ —

Today, Jeff is a middle-aged middle manager, who earns a middle income in a mid-sized corporation. He has a wife and three kids that he loves like crazy. He considers himself happy and successful, and by any conventional standard he is. He enjoys his work and rarely complains about anything to anyone. Jeff doesn't remember the story of his grandfather and the homework assignment, even though it produced some of the most powerful feelings he'd ever experienced.

But the feelings themselves, the horrified failure, the stirring and exhilaration of understanding, and the rush of glorious celebration, these feelings wash through him in some form, to some degree, and in various combinations almost constantly.

Jeff still has a favorite place. It's a lot more grown up than the one he had when he was six, but it serves the same purpose: he goes there to be alone when he feels particularly upset or inspired. His wife would tell you he spends a lot of time there, and lately it's often without enough light to read the magazine on his lap. Actually, most of the time Jeff just thinks. He thinks about a question that, as he has gotten older, has begun to claim more and more of his attention. He thinks about why so many things don't work the way they're supposed to, and why other things that seem to be working okay go horribly wrong.

When the space shuttle *Challenger* exploded in 1986, Jeff first heard about it on his car radio driving home, and he followed the investigation for months afterward. What he found most depressing was that the circumstances leading to the explosion reminded him an awful lot of certain aspects of his own job.

Jeff learned that some engineers had tried to warn NASA about the chance of an explosion and had actually tried to stop the launch. In fact, they'd been trying to point out a certain design flaw for more than a year. Sometimes NASA and the people the engineers worked for seemed to listen, but they never did anything. After a while, the engineers were branded as chronic complainers and troublemakers, even though they were right and had evidence to prove it. Over time, all their warnings were either denied, rejected, ignored, or overruled. The engineers' employers and NASA just wanted to stay on schedule.

The engineers became more emphatic and warned of a real catastrophe if corrective action wasn't taken—it was like trying to punch holes in a big, soft balloon. NASA decided to launch anyway, and the fireball that ended the lives of six astronauts and one teacher was visible all the way from Jacksonville.

When the investigation began, the people who had approved the launch tried to cover up the facts. Instead of being rewarded for their honesty and integrity, the engineers who'd been raising the red flag became outcasts. They were criticized, passed over for promotions, and generally blackballed. It was a long time before the next shuttle went up, and Jeff made a point of not watching TV when it did.

Situations like this really got to Jeff. There seemed to be so many of them, and they seemed to hit closer and closer to home. In March 1989, an Exxon oil tanker, the Valdez, went aground off the coast of Alaska, dumping 1,250,000 barrels of oil into Prince William Sound. This time, the investigation showed that the technical inspector at the *Valdez* terminal had been screaming for attention since 1983, trying to warn of an inevitable catastrophic spill, but nobody listened.

The background was all too familiar. Due to budget cuts, Alaska had reduced its program to supervise and inspect the big oil pipeline and hadn't monitored oil operations closely since the 1970s. They performed clean-up tests at sea with a simulated 60-barrel spill, using oranges instead of oil, and still barely passed. In an effort to show the danger of even a small spill, the inspector videotaped the test. When the local oil company consortium found out, they banned cameras from all future tests, openly referred to the inspector as a troublemaker, and tried to have him fired. The inspector, still fighting to be believed, insisted that a catastrophe was imminent. When the *Valdez* spill occurred, the consortium sidestepped the issue and blamed Exxon, which still bears most of the responsibility. Afterward, the inspector was demoted and transferred to another site.

And again … just a month after the *Valdez* oil spill, a gun turret on the battleship USS *Iowa* exploded, killing 47 sailors. All possible witnesses died in the explosion. The Navy spent $4,000,000 investigating the accident, conducted 20,000 tests, and concluded that no technical cause could be found to explain the explosion. In their final report, the Navy concluded that a gunner's mate had set off the explosion deliberately because a shipmate had rejected his homosexual advances. The families of both sailors vehemently protested the reports, but the naval investigation was closed.

Parallel federal investigations accused the Navy of using evidence selectively and found flaws in their inquiry, the technical tests, the structure of the investigation, and the psychological analysis. A new series of tests found evidence that the explosion was an accident and enabled investigators to reproduce it experimentally.

In the absence of eyewitnesses, the victims' families collected evidence from their sons' and husbands' letters and from conversations that occurred before the explosion. This material presented clear indications of unsafe storage and management of explosives, pointing to a high risk of explosion in the chronically overheated turret. Many months passed before the alleged saboteur was vindicated.

Jeff could imagine himself as one of those sailors, seeing the problem and knowing how to correct it but afraid to say anything criticizing his superiors. He might be transferred to some horrible duty, put on report, or even downgraded. He knew how fear worked and what happened to whistle-blowers.

Jeff attributed this recent chain of disasters, which he felt began with the Three-Mile Island nuclear plant accident in 1979, to a general deterioration of quality in the late 20th century. He figured it was a sort of critical mass of incompetence that just happened when machines and control systems got too complicated for people to manage safely, even though they kept trying to make everything faster, better, and cheaper. He began to think that people were just too greedy or stupid and that there was nothing anybody could do about it.

One day, Jeff saw a movie on TV about the Japanese attack on Pearl Harbor. That happened before Jeff was born, before computers, before nuclear power, and before color TV and paper shredders and voice mail. Jeff wasn't much of a history buff, but after a few minutes he was riveted to the set with a sick and depressing fascination.

December 7, 1941. Almost one year earlier, the US ambassador to Japan had alerted the State Department to rumors about a possible Japanese attack on Pearl Harbor. Most reports were ignored and not communicated upward. Eight hours before the attack, Army intelligence intercepted a message from Tokyo to the Japanese embassy in Washington, ordering the staff to cancel all meetings and destroy their code machine. Immediately, the general in command ordered a warning to be sent to Army headquarters in Hawaii by the fastest means possible. Instead of phoning Hawaiian headquarters, the communications department sent the general's message by telex—it arrived the day after the attack.

Four hours before the attack, an American destroyer sank a Japanese submarine in Pearl Harbor Bay, but the Navy didn't notify the Army. One hour before the attack, an Army radar operator tracked a large fleet of planes and reported it to headquarters. The officer who received the report assumed the planes must be friendly and did nothing.

At the instant of the attack, those watching with their own eyes still denied its reality. As the bombs began to fall, people thought it must be a case of malfunctioning bomb-release mechanisms, crazy pilots, or simply a very realistic training exercise.

Jeff watched the whole movie with an eerie feeling of disbelief. It was the same story, again. This meant his idea of late 20th century high-technology overload was wrong. The problem was older, a lot older, and Jeff didn't understand it at all. If it wasn't high-tech burnout, what was it?

It didn't answer this question, but the last line of dialogue made Jeff feel a lot better. It was spoken by the Japanese admiral who had commanded the attack. This man had serious doubts about a war with America, but being a dutiful Japanese officer, he planned the attack anyway. When it was over, and the pilots radioed their success back to the flagship, everyone in the fleet celebrated, except for this admiral. He just looked sad and said, "I fear all we have done is to awaken a sleeping giant, and fill him with a terrible resolve." When Jeff heard this, his distress was replaced with pride. He knew the admiral had been right.

Then Jeff started to think of other things that had happened over the years, things that were as big as the tragedies that had been occupying his mind so much but that were great things, wonderful things. He thought about *Apollo 11* landing on the moon in 1969, a feat that had always amazed him, having grown out of a space program that was a mere 10 years old. He thought of the US Olympic hockey team in 1980 that had won the gold medal against impossible odds after having been the acknowledged underdog for months. He thought about winning World War II, fighting on two fronts on opposite sides of the earth, gearing up to defeat the world's most fearsome armies in less than five years.

Jeff's thoughts jumped to other success stories that he'd read about over the years, stories that were much closer to his own life and work. He thought about the championship teams at Federal Express, Harley Davidson, Xerox, and Johnson & Johnson, whose successes gave him the same feeling of celebration as that first footstep on the moon, or the flood of blue-clad figures on the ice at Lake Placid. There were even some bright spots in his own career, things he hadn't seen in this light before.

Last fall, he threw together a sales presentation in an impossible time frame, winning him and his firm an important new client. And then there was the time a few years ago when he helped develop a new marketing system, which rapidly led his company to create a new line of business and won him a major promotion. His thoughts carried him back in time, and an image of his grandfather floated up in his mind, followed by the face of a teacher he once had as a kid. A jumbled flood of memories came to him: baseball games in high school, building a garage with his father, getting his first job. He started to feel good, better than he'd felt in weeks, and he didn't know why.

Jeff was feeling the birth of a tremendous insight that was almost within reach, but he couldn't quite see it. He felt sure it was a wonderful, beautiful thing—something that would waken the sleeping giant inside him and anyone else who could understand it. He knew it had something to do with trust, pressure, and not being afraid.

His boss was currently on a kick about "doing it right the first time," and Jeff felt that this vague, half-formed perception might actually be the thing that would let them do it. If he could just work it out clearly, bring it to his job in a usable form, he was sure his company would rocket above their competition. And if his current mood was any indication, they would all feel great in the process.

Jeff felt 10 feet tall. He put on some Charlie Parker, opened one of his son's encyclopedias, and spent the next three hours reading about Cape Canaveral, Neil Armstrong, and footprints in the Sea of Tranquility.

— ✧ —

Jeff never figured out the rest of the puzzle. He didn't bring the idea to his boss, because it never became clear enough to articulate. His boss still coaxes him and his staff to do it right the first time, but somehow they very rarely do.

The real key to the puzzle is wrapped in a slice of irony that Jeff wouldn't have guessed in a million years. The cruel, shameful tragedies of Pearl Harbor, the *Challenger,* and the Exxon *Valdez,* and the sweet, dreamlike victories of *Apollo 11,* the Olympic hockey team, and the Harley Davidson comeback are all products of the same thing. They are all fallout from exactly the same current of American culture, all consequences of something that lies so deep inside us we don't know it's there.

Jeff would be amazed to know the truth—that the twin seeds of defeat and victory lie at the heart of every American in a childhood story of initial failure followed by careful support and triumphant celebration. He'll always do his best, and he'll always try very hard, but he'll never do it right the first time. That's the other secret.

It's too bad Jeff and his boss didn't understand the simple truth, but unfortunately, they are too focused on trying to do it right the first time to see that it is the second time that really counts in America.

Americans don't do it right the first time; they just don't. But when that second time comes around, not even the stars are out of reach.

Memory is a child walking along a seashore. You never can tell what small pebble it will pick up and store away among its treasured things.

—Pierce Harris

— ✧ —

Trust your hunches. They're usually based on facts filed away just below the conscious level.

—Joyce Brothers

— ✧ —

Civilization no longer needs to open up wilderness; it needs wilderness to open up the still largely unexplored human mind.

—David Rains Wallace

1 MARILYN'S LAND OF DREAMS

THE RENAISSANCE

Quality, not surprisingly, is something I think about a lot.
When I began my assignment as AT&T Quality Manager, I knew that *quality* as a stand-alone idea was nothing new in American business. As a concept, a goal, or an item for general discussion, it had been turning up on corporate meeting agendas for decades. I also knew that quality was nothing new at AT&T. In fact, it occasionally strikes me that the real birth of American business quality occurred in 1925, with the formation of AT&T's Bell Telephone Laboratories. This was perhaps the first business institution created specifically to develop new products and technologies of higher value and performance.

For the next 50 years, the Bell System thrived, and through the quality of its communications network and many unique products and services, it earned the respect of millions of customers. The Bell System was a rare and wonderful machine: Bell Labs developed new products, AT&T's Western Electric manufacturing unit put them into production, and, as a rule, customers were more than satisfied. Ma Bell's standing as a monopoly didn't seem to bother anyone. America had the finest phone system in the world, and with the passage of time it seemed to grow consistently better.

Economic, regulatory, and technological changes in the decade before divestiture gave rise to an AT&T that was somewhat less secure in the marketplace. Although AT&T still produced high-quality products and services, it was often at a correspondingly high cost; but the size and stability that AT&T had achieved made it hard for the organization to evolve or react to changing conditions. It was something of a juggernaut, rolling along with familiar habits, methods, and procedures. The steady revenue from operating the only phone system in the country tended to minimize any concerns about efficiency or competitiveness.

In 1984, the long monopoly ended in a complex, painful divestiture from which all of us are still recovering. The Regional Bell Operating Companies, the so-called "Baby Bells," became independent and began their own marketing, sales, and support activities for telephone service in their own areas. What remained of AT&T faced a growing number of competitors, aggressively expanding into consumer products, information services, even long distance networks. As AT&T reeled from these impacts, I imagined a giant grown soft and complacent from years of easy living suddenly cast into the wild to be harried by wolves. It was a grim picture.

Wolves notwithstanding, AT&T moved forward. The year after divestiture, it formed a new entity called Network Systems, which combined most of Western Electric, still AT&T's manufacturing business, with their corresponding Bell Laboratories research and development organizations. Network Systems involved over 100,000 people and many diverse business functions, including a small band of dedicated hopefuls known as "the quality group." As the preface describes, I moved into this group late in 1985, working for Quality Director Ray Peterson.

This group had a complex and somewhat daunting mission: to create a Network Systems quality renaissance. They had developed a plan to implement certain quality tools and methods, proven for years in design and manufacturing, in all other business functions: operations, sales, finance, marketing, and service. They all believed (as I came to believe, very quickly) that quality was the key to AT&T's future in the global marketplace and intended to prove to a sizable crowd of disbelievers that quality saves money rather than costs money.

Although Network Systems executives endorsed the renaissance project, I found implementation to be far more difficult than anticipated. Many of our efforts were similar to those that had been hugely successful in Japan, but for no reason that I could see, they rarely worked here. The results were spotty at best and were seldom sustained for long periods. In organizations where people had never thought of quality practices as necessary anywhere beyond the factory floor, I confronted what seemed to be impenetrable barriers to both understanding and action. I kept running into an attitude that was something like *Real Men Don't Do Quality:* "It may be fine for those assembly-line types in manufacturing, but it's not for us folks in marketing." When I learned our Quality Leadership Conference, the virtual flagship of the renaissance project, was to be put on hold indefinitely, I felt angry, confused, even a little betrayed, but nonetheless determined to keep fighting.

What was the problem? It was clear enough that AT&T was caught up in a national quality emergency; but why were we having so much trouble getting people to take action, to do something about it? I realized that getting around whatever deep-seated roadblock we kept hitting would take more patience and a new approach.

In my search for a new angle on this problem, I thought back to everything I'd ever experienced, read, or heard concerning quality issues in business. As the preface tells, this internal stock-taking eventually called to mind Dr. G. Clotaire Rapaille, with whom I'd worked a few years earlier. Now, as I ran into one quality roadblock after another, I thought of his work again and suggested to Ray that we spend some time investigating American cultural characteristics. I felt that Dr. Rapaille's methods might reveal the aspects of our culture that either tend to support the approaches so successful in Japan or tend to inhibit them. The results might move the renaissance project past its impasse.

THE HEART OF THE MATTER

My next contact with Dr. Rapaille was like the sliding of a few small stones that starts an avalanche; it set many things in motion, some of which haven't stopped yet. In a series of phone calls and meetings that grew progressively longer, we began spending more and more time discussing Network Systems quality issues, and it wasn't long before we were calling each other Clotaire and Marilyn. During the last seven years, he has become a close friend and a priceless ally in my ongoing pursuit of quality for AT&T. What follows in this section are the basics I learned from him in 1986, elements drawn from psychology, anthropology, and biology. These elements represent the essentials of his work as I came to understand it.

Clotaire's work focuses on the transitional years of early childhood, roughly between the ages of one and six. During these years, children are shaped by the forces and people around them and begin to take on the qualities of the men and women they gradually become. For me, some of the most surprising and valuable information Clotaire had to offer concerned the learning of language during this period.

I learned that as people learn a particular language, they acquire a distinctive set of tools to relate, analyze, and function in the world. Language lets people grasp the meaning of objects, actions, or qualities and the relationships among them. Learning these relationships is not passive, experienced simply as information received by the mind. It's an active process, relying on personal discovery and grounded in emotion. Indeed, without emotions, people might be incapable of learning a language effectively.

People learn individual words through specific emotional experiences. Such experiences are not necessarily the first time they've heard a certain word, nor even the first time they're able to use it correctly. A child may remember the sound of a word first and may be able to reproduce it. Later, the same child may learn at the simplest level what that sound represents: a tree, a dog, or a car. But often the deepest learning of a word, the incorporation of all its flavors, connotations, and implications, occurs when the learning experience is surrounded with strong emotion. Such an experience creates an emotional imprint in the child's mind, an imprint that carries the word's most profound meaning. This imprint is like a pathway carved and shaped by emotions and carries a permanent link to a particular word. After an imprinting experience, people will use this same pathway for years, perhaps for the rest of their lives, every time they relate to whatever the word describes.

To this day, I'm struck by a certain irony in this process: as people grow, they lose track of the original imprint. The original experience and its aura of powerful emotions are usually forgotten and remain stored only in the unconscious. Consciously, people remember only the label that comes to represent those emotions: the word.

I recall one of Clotaire's stories: a scenario for a possible imprint of the word *glass*, and the related concepts of fragility and breaking.

Once upon a time, when you were very small, you were drinking from a glass. Your mother told you to be careful, because if you dropped the glass it might break. You had no idea what she meant, because you'd never seen a glass break. Now generally, you did what your mother told you to do, because you wanted to please her and of course were completely dependent on her. At this point in your life, the world was a confusing and often threatening place, and mom was one of the few things you could count on. With good reason, you listened to her. But at the same time you were curious. Like all human beings, you explored, tested limits, and took risks. So, trying not to be too obvious about it, you let the glass slip from your grasp and fall to the floor. It broke.

Immediately your mother became very upset, maybe even angry, and you experienced a strong and unpleasant reaction to her emotional state. She yelled at you but calmed down quickly enough. She didn't hurt you. In fact, she continued to feed you, take care of you, and love you. You also noticed that your mother could not reassemble the pieces of broken glass—the change in its physical state was permanent.

Suddenly, you were assaulted by a stressful situation involving your mother's anger, your own fear and anxiety, the loud noise when the glass shattered, and the broken pieces on the floor. All these elements gave you a complex emotional imprint related to the concepts of *glass*, *breaking*, and *fragility*, and possibly others as well. This experience may have occurred before you could speak at all, or it may have happened when you were four and had been talking about glasses for some time. But until you've had this experience or one like it, you haven't been imprinted with the word's deepest meaning. After the imprint experience, the link between the word and the emotion of the experience is permanent—you may forget the experience, but you'll never lose the imprint.

Consider the following analogy to explain the nature of an imprint and its effect on our minds. Picture a flat desert landscape—the ground is dry, hard, and featureless. When it rains, the water pools here and there but has little effect on the ground; the ground is too hard. But when a real storm comes and water pours down like a flood, the ground starts to erode. The rush of water begins to create small channels, then gullies, ditches, riverbeds, and perhaps even canyons, depending on the power and duration of the flood. Afterward, the ground has been permanently changed, engraved with a critical pathway that reflects the exact conditions of wind and rain during the storm. From now on, whenever it rains the water will flow through those same channels, making them deeper and wider.

This is how people are imprinted by emotions. With a particular experience comes a flood of emotional energy that imprints a certain pathway in the mind. From then on, whenever people have a similar experience, their emotions flow through the same channels and the imprint becomes deeper. Each repeated experience reinforces the imprint, validating the emotions and strengthening their links with certain words.

An *imprint structure* is the mind's permanent record of the imprinting experience—the complex sensory, emotional, and verbal episode converted to a form the mind can store and reuse. Think of those channels and riverbeds eroded in the desert. They exist, and will continue to exist, whether or not any water flows in them. They were created by a flood, but when the flood is over, the structure of channels and riverbeds remains. This is the nature of an imprint structure: it's the riverbed, not the river. After a flood of emotion has passed, the imprint structure remains in place, full of potential for later experiences.

In his book *The Fifth Discipline*, Peter Senge sheds considerable light on this phenomenon, discussing language and its role in programming the subconscious (Clotaire focuses on the imprinting of the *unconscious*). Senge points out: "There are many ways by which the subconscious gets programmed. Cultures program the subconscious ... Perhaps most subtly, language programs the subconscious.

The effects of language are especially subtle because language appears not so much to affect the content of the subconscious, but the way the subconscious organizes and structures the content it holds."[1]

ARCHETYPES

Clotaire's work indicated that the structure of any given imprint is very much a function of the culture in which the imprinting occurs. For this reason, people sharing the same culture generally share very similar and often identical imprinted structures for many words. This is another important characteristic of imprint structures: they tend to be the same for virtually all members of a given culture. Regardless of the tremendous variation among different imprinting experiences, the *imprint structure* for a word in a particular culture most often takes just one form.

Imprinting experiences are rich with detail: emotion, language, and sensory input. But the imprint structure is independent of all this detail. To better understand the difference between the structure and the experience that creates it, consider the first time you hear a particular song, maybe on the radio in your car—such an experience is full of detail. Maybe it's a warm summer night, so you have the window open and can smell food from a nearby restaurant. In the middle of the song, a police car goes by with lights flashing and siren wailing. The radio station isn't coming in too well, and there's a lot of static; but you like the song so you leave it on anyway. All these perceptions are part of the experience of hearing that song for the first time; but these perceptions are not the song.

The next time you hear the same song, it's played by a jazz band at someone's wedding reception. And the next time, it's played by a string quartet on a television program. And the time you hear it after that, it's part of the background music in a supermarket, interrupted by announcements for discount specials in the produce aisle. Each of these experiences of hearing the song is very distinctive. They all have their own sets of perceptions, but in each case you recognize the song immediately. This is because the

---

[1]Senge, *Fifth Discipline,* p. 366.

structure of the song is independent of the perceptions and details of the individual experiences.

To recognize the song after hearing it once, your mind doesn't need all the perceptions that were present that first time; in fact, your mind often doesn't even need the melody. To remember most songs you need only their most basic structure: the rhythm and the timing of the notes. This simple structure enables you to reexperience a song whenever you hear it again, no matter what the circumstances or the musical arrangement. While each person has a different sort of experience hearing a song for the first time, full of individual perceptions, the basic musical structure that's stored in memory is the same for everyone, regardless of where, when, or how they heard it.

As with music, imprints have a particular structure, and Clotaire's theory contends that those who share a culture share common structures. He calls this unconscious imprint structure a *cultural archetype*™ (är′ kə tīp′). The dictionary defines an archetype as an original pattern or model; prototype.

The eminent Swiss psychologist Carl G. Jung was the first to describe archetypes. Jung's archetypes are universal—he believed all human beings are born with them and therefore are possessed of certain instinctive reactions that are biologically prestructured at birth.

Working from a foundation of Jung's universal archetypes and Sigmund Freud's ideas of the individual unconscious, Clotaire developed the concepts of the *cultural unconscious*™ and cultural archetypes; he also developed research methods to identify archetypes associated with particular words or ideas. Hereafter, whenever we use the word *archetype*, we refer to Clotaire's cultural archetype. When we use the phrase *cultural study*, we refer to AT&T's cultural archetype study of quality.

The deeper I looked into Clotaire's work, the more I realized that archetypes are slippery, elusive things. They can't be observed directly, nor can they be revealed by standard techniques of behavioral analysis. The behavior surrounding a certain thing may suggest the corresponding archetype, but the behavior alone won't convey

---

cultural archetype and cultural unconscious are registered trademarks of Archetype Studies.

much about what the archetype really is. A behavioral study cannot reveal the imprint structure produced by a complex tapestry of emotions—it may identify a few isolated emotions but cannot indicate the entire structure of an archetype, if in fact one exits at all. Not all words are connected to specific cultural imprint structures.

Consider the example of the word *glass*. If you stroll into a drinking establishment and observe lots of people handling glasses, you might have some degree of success discovering their collective attitude about them. You would notice that people handle glasses with a certain amount of care, more care than one might exercise with, say, a book, a stapler, or a basketball. You might see that people are quite careful about how they hold glasses and are fairly gentle when setting one down on the bar. But that's about it.

Watching people's behavior with drinking glasses, you learn very little of the archetype behind the word. You see no indication of the confusion, shock, anxiety, or embarrassment that contributed to the imprint for *glass* and its related concepts of breaking and fragility, even though that imprint almost certainly exists inside everyone you see. People's behavior, especially with adults, reveals very little of the cultural unconscious. To reveal an archetype, you must study its imprint directly.

Consciously, people aren't aware of their own imprint structures and often don't even remember the experiences that engraved these structures in their minds. Awareness of emotional imprints or archetypes isn't necessary for learning a language—all that people need to remember consciously are the words with which imprints are linked. Nonetheless, imprint structures are a permanent part of everyone. They often influence people's behavior and affect their decisions. Once you become aware of a certain archetype, it seems as though people relive and repeat the original experience without being aware of it. Whenever someone knocks over a glass, he or she will always lunge across the table to grab it, whether it makes much sense or not. In this way, people reinforce and perpetuate their archetypes.

Studying archetypes with Clotaire's methods can help identify the entire structure that's been etched in the mind and, most important, the emotions that created a particular imprint. Often in adult life, people try to motivate, influence, or encourage others to behave in certain ways or to make certain choices. When these efforts succeed, the messages and images are usually working in accord with the appropriate archetypes. When these efforts fail, the messages and images are very likely conflicting with the archetypes.

Consider another example from Clotaire's work, a study he conducted on the archetype of doors; we're talking just basic doors here, front doors and back doors. This study revealed that the imprint structure of *door* for Americans is not just any good, safe door, but a very specific kind. It's not one that keeps bad things from getting in, but instead is one that ensures people can easily escape if danger threatens. For Americans, the archetype of a door is a way out, a way to safety.

By contrast, a safe door for the French is one that keeps danger from entering. In France the safest door is a drawbridge in a thick stone wall, preferably behind a moat. Guided by these unexpected findings, an American door manufacturer altered its advertising, promotional materials, even the way its sales people demonstrated doors to customers. In time, they successfully edged ahead of their competition in the US market.

A PRODUCT OF OUR CULTURE

Clotaire contends that archetypes are basic patterns for all aspects of human life and can vary enormously from one culture to the next. From an early age, archetypes begin to organize and influence many of the perceptions that we develop as adults.

Archetypes exist in the mind at a deeper level than most traditional market research can look. The pyramid graphic, shown in Figure 1.1, makes it clear just how deep this is. The different layers in this graphic correspond to the different depths at which information exists in the human psyche.

At the very top is what is known to people consciously, and also what is generally visible to others; it's the smallest part of the whole and the part most easily influenced by opinion.

Below this conscious level lies information that is not available to people directly: the individual unconscious. While it forms more of each person than the conscious self, even the individual unconscious represents a relatively small part of the whole. Although not

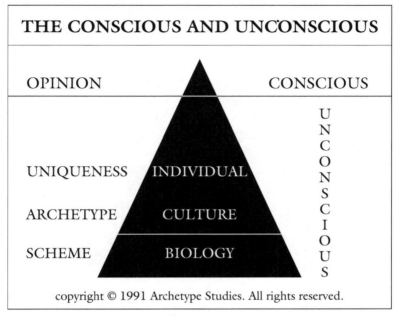

Figure 1.1

available to us consciously, information stored here is still unique to each person. It comes from our upbringing, our families, and our environment.

Below the individual unconscious is the cultural unconscious— information held inside people at this level is a product of their culture. It's this area that one explores when searching for archetypes. At this level, everyone loses a certain degree of individuality, because these imprints and patterns are essentially universal within a given culture.

At the deepest level in the pyramid lies the biological self, the schematic human being that comprises characteristics that are identical for everyone. Biologically, all people are very much the same, sharing the same basic anatomy, physiology, and genetic codes. And they tend to stay the same from one generation to the next: people have carried the same DNA for millennia, as long as there have been *Homo sapiens* on earth.

The rate of change in people's lives has a special relationship to the pyramid model and varies as you move down through the four levels. At the top, the conscious level, things change with great frequency, but as you move toward the bottom, the rate of change decreases, until you eventually reach a point where there's almost no change at all.

Most market research explores experiences, memories, beliefs, and thoughts that are available to people at the conscious level: what they think and how they feel. In this type of research, promising candidates fill out questionnaires, discuss new products in one-on-one interviews, or engage in lengthy discussions in focus groups. This approach can reveal much about what people believe is true based primarily on their knowledge and opinions. But knowledge and opinions, even when clear and honest, tend to change quite often. What people believed to be true last year could be viewed as utterly false this year.

Now consider the cultural unconscious. Things can change at this level, but, as Clotaire often says, they change at a glacial pace; a given archetype may change noticeably in the course of a few centuries but certainly not in the average lifetime. This level is characterized by tremendous stability, and this is what makes archetype studies so valuable. Once you identify an archetype, you can rely on it for many years. This makes a well-crafted archetype study very effective as the foundation for a long-term business, marketing, or motivational strategy.

Archetype studies yield information from a place deep inside the mind. They reveal the logic behind people's emotional reactions. They are more efficient and cost-effective than traditional market research, since fewer study groups are needed and the study cycle is shorter. And because they yield information that is stable within a culture for a long period of time, archetype studies can often predict human behavior with a high degree of accuracy.

As we investigated Clotaire's archetype work in other countries, Ray and I found many success stories. Some companies had commissioned archetype studies to help predict the success or failure of domestic products and services, before their introduction in a foreign culture. In many of these cases, marketing and advertising campaigns were completely redesigned based on information the study revealed. Other companies had conducted archetype studies to explain why a product that was successful in a domestic market failed abysmally in another country. When no other investigation produced an answer, the archetype study not only explained why the failure occurred but also suggested how to turn the failure into a success.

During the last 17 years, Clotaire has conducted archetype studies around the world. Their results are often held as proprietary information by the company or organization that commissioned the study. These studies can be applied to general categories, like cars or cosmetics, but in some cases can also focus on individual products. Applied to a specific brand of soft drink, for example, such a study might reveal information that can ensure customers' product loyalty.

Archetype study findings have affected important decisions in a wide range of circumstances. They helped France determine its nuclear power strategy and showed certain American companies how to market new bakery products in Europe. They have helped beer, soup, and cosmetic companies market their products in Japan and have persuaded a number of banks to design their physical, social, and information systems to serve different ethnic communities in several different countries.

After spending several months investigating Clotaire's work, I had become convinced that archetype studies are quite powerful. He had demonstrated time and again that such studies can provide businesses with some of the most effective information available, information on products and services, customers and employees, our individual past and our collective future.

As part of the Network Systems quality group, my interests lay along slightly different lines. I wasn't interested in selling products or services, or determining the roots of customers' loyalty to AT&T. Ray and I wanted to uncover something even more elusive, if in fact it existed at all: the American archetype of quality. If doors turned out to be avenues of escape, what would lie at the heart of quality?

THE STUDY

In early 1986, I began working with Clotaire to lay the foundation for an AT&T archetype study of quality. The study was ultimately conducted between February and October 1986. As a source of cultural information, it used groups of people that represented a cross-section of American culture. Participants were chosen according to income, age, race, cultural background, religion, and sex, so that the final study group constituted a representative sample. Participants were also distributed geographically across the United States. But before these people could be selected, the quality group had to address a particularly troublesome question. We knew that an American quality archetype would apply to all Americans. But for the purposes of the study, who or what is an American?

Native Americans notwithstanding, America is largely a nation of immigrants. The people who have been coming here voluntarily for the past 300 years or so are attracted to this country on different levels: they're drawn by many visible American opportunities, social or economic, but they are also drawn unconsciously by a wide range of distinctly American archetypes.

In the United States, when immigrants cling to their original culture and form their own communities in major cities, it might seem to be a rejection of American culture, a defense to wall out any influence of our language or customs. But remember that behavior is usually not an indication of an archetype. Often, when immigrants come here, they are drawn by that uniquely American dream, that part of our culture that purports to embrace all people equally and promises them the freedom to live as they wish. When Puerto Ricans or Ukrainians or Koreans form their own communities, with all the freedoms guaranteed by the constitution, they are not rejecting the American archetype; they are reveling in it. It's often in such communities as these that American values are the strongest.

This situation makes the study of cultural archetypes rather challenging, particularly in the United States. Here we have two very different types of citizens: people that embody American archetypes because their families have lived here for generations and immigrants who embrace American archetypes because they were attracted by them from afar. Then, of course, there are the huge numbers of second- and third-generation immigrants that lie somewhere between the two extremes. To cope with this sort of diversity, Clotaire had developed specific criteria to qualify candidates for study. These criteria enabled him to select people not only who represented a particular culture but also whose imprinting had not been blurred by influences from other cultures. Here are the basics: an individual must have spent the first 15 years of life within the culture being studied, must have been born to parents who were members of that same culture, and must have spoken mostly the culture's native language in the home. These requirements ensured that a study's results would reflect characteristics unique to a particular culture.

For the Network Systems cultural study, Clotaire designed a series of study sessions, each one broken down into three parts. Because I met all the criteria, I served as a member of the study group for the first session.

The session I attended was conducted in a hotel conference room during the evening. The group was quite informal, involving about 20 people between the ages of 20 and 70. On average, about half of each group had been drawn from the ranks of AT&T employees.

The first part of the evening was a simple, informal discussion during which we offered our thoughts and opinions about quality. What did quality mean to us? What were all the different ways we felt about quality? How did the perception of quality change if we imagined our roles switching from being a producer of quality products and services to being a consumer? We continued this discussion with Clotaire for about 45 minutes. It helped us focus consciously on quality issues.

The second part of the session was a series of structured free-association exercises. I understood much later that Clotaire uses these exercises to discover the feelings and concepts that lie behind the words being studied. Because language plays such an important part in imprinting, he believes word associations can begin pointing to the unconscious latent structure of a particular word. This was another unexpected discovery for me—that many words conceal some sort of latent structure.

Lurking behind a word's accepted meaning, there are often subtle flavorings, connotations that aren't so obvious. These indirect meanings or implications can reveal latent structures. For example, compare a business organization chart with a military chain-of-command; the two organizational structures are virtually identical. This similarity appears just as strongly in the structure of language. In modern business, there are *general* managers, *strategic* planning, and an entire vocabulary that shows a strong latent structure behind it: the military. For another example, look at the image of the woman shown in Figure 1.2. When you see this drawing, you'll first see either an old woman or a young woman. Don't be surprised if you can only see one or the other—it's often very difficult for your mind to see the second image once you've recognized the first. (*Hint:* the old woman's left eye is the young woman's left ear.) You might say that behind the old woman is the shadow of a young woman and vice versa. Each image contains the other as a latent structure.[2] The word associations in the quality archetype study worked along similar lines and enabled Clotaire to discover some of the latent structures underlying quality.

THE YOUNG GIRL AND OLD WOMAN

Figure 1.2

---

[2]Edwin G. Boring brought the concept of the Old Woman—Young Woman to the attention of psychologists in 1930. The picture was created by cartoonist W. E. Hill and published in *PUCK* in 1915, entitled "My Wife and My Mother-in-Law."

The third part of the evening was the most significant. Clotaire guided us into a state of relaxation and asked us to reexperience, as vividly as possible, the very first time we learned the concept of quality. He then led us on to explore quality at other significant moments in our lives. Before any further discussion, he asked us to write down the quality experiences that seemed to be the most meaningful. He encouraged us to be as descriptive as possible, to include all objects, people, feelings, timing, settings, and any other salient details. Together with the word-association exercises, these writings were analyzed for patterns and common themes. It was in this material that Clotaire hoped to find a single, unifying structure for quality.

When the session was officially over, we were struck by an unexpected and delightful phenomenon: the participants wouldn't leave. This score of strangers who had donated their evening, answered our questions, written down their stories, and generally mused about quality had become fascinated by what we were doing. The process we took them through—the word associations, the guided relaxation, and the exploration of childhood—was like nothing they'd ever experienced. Now they were full of questions of their own: Why were we doing this? Whose idea was this? What did we expect to learn? And what had we learned that evening? Many of the participants stayed with us at the hotel for hours, talking about their stories, thoroughly intrigued by the experience.

Clotaire, the study team, and I had plenty of questions of our own. What did a particular word association mean? What was the significance of a certain person's first quality recollection? Were there any common themes among the stories? To us, these scraps of paper with words and stories were like moon rocks: priceless, fascinating, and utterly unique. We were all tremendously excited by the study and wanted desperately to understand the information as thoroughly and quickly as possible. Each session was a powerful learning experience, full of discovery and exploration, for me and all those involved.

After each set of three sessions, Clotaire, the study team, and I spent up to two days exploring the substance and significance of the stories. Perhaps I can best describe this particular experience by comparing it to panning for gold or sifting through jigsaw puzzle pieces, looking for any that might fit together into a meaningful shape. At times, we would turn up interesting or provocative clues in the data, which would often prompt a slight shift in our focus for the next session. This might lead to new avenues of exploration for subsequent evenings or variations on the exercises that might shed light on questions we hadn't answered.

There was no sudden revelation. The archetype structure appeared gradually—a piece here, a fragment there. During this period, I often felt like a paleontologist, chipping flakes of stone from the bones of a dinosaur—little by little, the study sessions exposed more of the underlying structure. Common themes became more and more apparent, as memories, emotions, and other details from the stories accumulated along certain lines. After the seventh study group, the structure was complete. Several other sessions were held to confirm it, and each of these produced the same telltale pattern. This was it.

The study team and I were electrified. Ray had been very cautious about the study in the beginning and rather unsure of the whole venture. Well, he had become quite an archetype convert during the previous 10 months or so. He attended every study session. Whenever Clotaire was ready with feedback and analysis, Ray cleared his calendar (something that borders on the miraculous at AT&T) to accommodate Clotaire's schedule. At this point, after more than a year of false starts, dead ends, and many sleepless nights in pursuit of quality, I was more excited than a kid on the first day of summer vacation.

After learning the imprint structure, Clotaire, the study team, and I could accurately describe the first experience of quality for virtually all Americans, the experience described in different ways with many different plots and characters by nearly everyone in our study. Although the details of each person's experience varied, the basic story was always the same. This experience is Jeff's homework story, told in the prologue of this book. Looking even deeper, we realized that there was also a more basic pattern of quality, a true generic imprint story that applies to all Americans.

I had learned from Clotaire that when an archetype exists, its strength and power are directly proportional to the intensity of emotion surrounding the imprint moment. Well, the archetype of quality for Americans is strong, amazingly strong—the emotional intensity surrounding an American's first quality experience is staggering. Consequently, the first key finding of the study was that quality is very much an emotionally laden concept for Americans; it's not to be taken lightly.

Sometime later, when I read Tom Peters' book *Thriving on Chaos*, I realized the author must have an intuitive understanding of this important aspect of the quality archetype. He wrote: "I will soon turn to talk of systems and measurements, but it is essential to begin with emotion . . . Quality begins precisely with emotional attachments, no ifs, ands or buts."[3]

To my knowledge, Mr. Peters had not conducted any quality archetype research, but he knew what he was talking about.

---

[3]Peters, *Thriving*, p. 71.

If it weren't for the rocks in the bed, the stream would have no song.

—Carl Perkins

— ✧ —

The greatest obstacle to discovering the shape of the earth, the continents and the ocean was not ignorance but the illusion of knowledge.

—Daniel J. Boorstin

— ✧ —

Every time I ask a group of managers about their own crisis-provoked "miracles," there is no end to the stories they recount. Virtually every manager has seen it happen. But when the crisis passes, they almost always overlook the profound significance of what the crisis revealed.

—Robert H. Schaffer

2 SEEDS OF FAILURE AND SUCCESS

A FIELD OF FORCES

Wen Carl Jung first described the structure of a universal archetype in 1902, he depicted it as a field of forces created by the tension between points in opposition. Cultural archetypes, as well, can be depicted using a field of forces. Such a field of forces is not immediately apparent; as a rule, you can only see dim behavioral reflections of the tension that exists among the forces; the behavior may be apparent, but the source is not.

To identify every source of a person's behavior, you would have to examine every motivating factor in the conscious and unconscious. But to determine the source of an archetype, something having a strong impact on our behavior, you would look for a certain pattern, a distinctive sort of emotional logic surrounding a particular word. This would lead you to an examination of the memories, perceptions, and connotations of that word that grow directly from a given culture.

The AT&T Network Systems American quality archetype study revealed tension created by the opposing forces of actions and feelings. In the action area, tension exists between the states of doing what others expect or want and not doing what others expect or want—two opposite extremes. In the emotion area, the tension exists between the emotional states of feeling good about ourselves, having a strong sense of self-esteem, and feeling bad about ourselves, embarrassed. A graphic depiction of these relationships is shown in Figure 2.1. Jung called this four-cornered archetype structure a *quaternity*.

This remarkably simple diagram shows the states of emotion and action for an American's first learning of quality. Typically, Americans start out in their quality learning somewhere in the lower left quadrant: they don't do what others expect or want, and they feel bad about it. This corresponds to the point in Jeff's story when he didn't do his homework correctly and is terribly embarrassed when he realizes his mistakes. Fortunately, that's not the complete archetype. The diagonal arrow represents a transformation. It indicates a change from the time when people don't do what others expect or want and thus feel bad to the time when they do produce what others expect or want and thus feel good; this usually means

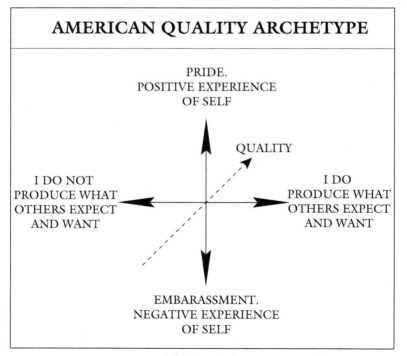

Figure 2.1

proud and satisfied. The success of this transformation depends on very specific roles being performed at certain times.

With this information alone, you can see that some of the messages often used to stimulate quality behavior conflict with the archetype and may actually produce the opposite response. Messages like "do it right the first time" and "zero defects" by their very nature are not consistent with the American quality archetype. Doing it right the first time and zero defects may be expectations, but as messages to Americans they're more debilitating than motivating; they make Americans feel controlled and restricted.

All Americans live with an internal, unconscious pattern, like a little voice in their heads. As much as they consciously want to do things right the first time, this pattern tells them that, more often than not, things are not done right the first time. When Americans are told to do it right the first time, an internal conflict arises, and they typically resolve it by taking fewer risks, playing it safe, and hiding minor mistakes as they occur. All Americans certainly want zero defects, but trying to inspire them with these catchy phrases or slogans goes against the imprinted package.

Another discovery from the study concerned the word perfection. In America, perfection isn't synonymous with quality but instead has unpleasant connotations. We were quite surprised at this.

For Americans, perfection is the end. Beyond perfection, there is nothing: perfection is death itself. If you reach perfection, there's nothing more to do. You've hit a dead end, game over.

Figure 2.2 shows another quality quaternity. The opposing forces here create tension between the beginning and the end of something, and between failing and succeeding. Remember, in the beginning, the imprint of quality is associated with failure. This is characterized as "only human." For Americans, this is what it means to be normal and to be associated with life—in the beginning, you're not expected to succeed. You fail, or at least stumble, and certainly don't achieve ultimate success; but that's okay, you're only human. If you succeed in the beginning, there's no glory for you and no reward. You didn't work hard enough.

This aspect of adult life reflects early childhood education. Many people received report cards that gave separate grades for effort and achievement. If you brought home a report card that

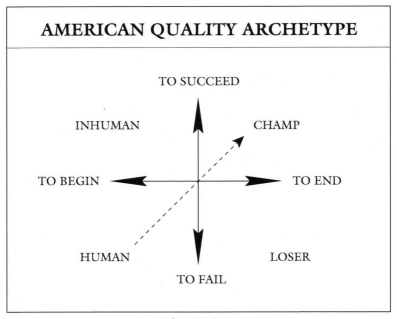

Figure 2.2

indicated high achievement as a result of high effort, shown in Figure 2.3A, you were proud, and your parents were proud too. You had worked hard and had succeeded very appropriately, the greatest effort for the highest grade.

If you brought home the report card in Figure 2.3B, your parents would react as though something were out of whack. They would probably think the work was too easy or that the teacher had marked the card unfairly, perhaps being too lenient in the achievement grading. High achievement with low effort doesn't feel right or make sense to Americans. If effort is low, achievement should be low. This report card is a signal that something is wrong. This attitude is also manifested in modern business; many managers express a desire for their people to work smarter and harder, to put both more thought and more effort into their work. But despite much lip service to working smarter, these same managers tend to reward those who visibly work harder. The employee who leaves work at 5 P.M. every day, because he or she is so efficient and well-organized, just doesn't seem to be doing as much as the one who's in

REPORT CARD	
ACHIEVEMENT	EFFORT
(A)	(1)
B	2
C	3
D	4
F	5

Figure 2.3A

REPORT CARD	
ACHIEVEMENT	EFFORT
(A)	1
B	2
C	3
D	4
F	(5)

Figure 2.3B

the office until 7 P.M. every night, sorting through heaps of paper-work with rolled-up sleeves.

Very early, Americans learn that for their achievements to be valued, they must put in maximum effort. They must at least look like they're working their hardest, even if it's just a smoke screen. In our culture, achievement with minimum effort will never bring the same accolades as when maximum effort is expended. Achievement without effort conveys arrogance or unfair superiority.

Back to Figure 2.2, our winner/loser quaternity. If you fail at the end, you're a loser and land in the lower right quadrant. Americans don't like losers, or losing at the end, but it's preferable to succeeding at the beginning. If you lose, you always get another chance. Consider bankruptcy; it gives you a chance to end a failure and start over. This characteristic lies very deep in the American archetype. Everyone who immigrates to America comes for a new start, a second chance, and they believe they will have an unlimited number of times to try.

Now consider the place where all Americans want to be: the upper right quadrant of our diagram, where people achieve success at the end of something. When you succeed at the right time in a project (when it's over), you're a champion, a winner. This is the classic story of the American hero. An American hero or heroine is someone who begins in the lower left quadrant, an underdog, with no advantage, no help, nothing. American champions overcome all adversity, withstand all pressures, and penetrate all barriers. They rise to their feet whenever they're knocked down, no matter how many times. American champions persevere against all odds, and finally, only in the very end, they win.

Every American athletic team that's ever had a big following started out as an underdog—the Dodgers, the Mets, the US hockey team at the Lake Placid Olympics in 1980. The members of this hockey team were classic underdogs. They had never played together before, were relatively inexperienced as a team, and had no hope of winning. They spent a mere six months training together. They weren't star players, just good team players; but they won the gold medal.

Their magic came from their shared dream. It was a magic conjured up by people coming together out of nowhere with no advantage and working intensely, fighting against all odds, against a professional Russian team, overcoming every setback but winning in the end. They became heroes and inspired the entire nation. This story is the archetype, the timeless story of success in America.

Many successful American movies tell the story of an underdog who becomes a champion and reflect the basic pattern of the archetype. If Rocky had knocked out Apollo Creed in the first round, few would have been interested in the story, and the film would have vanished shortly after release. The classic American success story is what provides this film's energy. It appears again in *The Karate Kid*, a movie that captured the hearts of Americans and inspired two sequels. These films work because they're built on the

American archetype. The people who wrote them may not have been aware of this fact, but they intuitively knew the pattern. They knew what would work for Americans.

Besides showing why Americans react favorably to certain stories, the archetype study also sheds light on the American emotional reactions to certain words. The study revealed that words often used in conjunction with quality stimulate negative feelings, very much like the word perfection. Words like standards, specifications, control, and maintenance do not motivate Americans. They actually shut down people's energy rather than feed their hunger for new possibilities. They make Americans feel controlled rather than empowered. All these concepts are important, but they must be put into perspective. Certainly, Americans must meet appropriate standards, conform to customers' specifications, control their processes, and maintain their equipment, operations, and software. But when work is described using these or similar words, Americans are rarely motivated to do it with the spirit or enthusiasm needed to be No. 1.

The study also found that some words stimulate very strong, positive feelings. New, change, possibility, opportunity, and breakthrough are all words that excite us. Americans associate them with creativity and innovation. Something new, something different, an adventure, a change, finding untapped resources, exploring unknown territory—these are the concepts that excite, energize, and motivate Americans. It was feelings like these that sustained the archetype study itself and gave the Network Systems quality team the energy to push forward despite the difficulties. They felt like they were breaking new ground, and, exhausted or not, they loved it.

STARRING ROLES

For virtually all Americans, the beginning of the imprint experience for quality was a time when they did not produce what others expected or wanted and thus felt bad. Their self-esteem was lowered—they were embarrassed. There's good news about this painful beginning: a successful completion results from transforming this negative emotion into positive energy, so the pain felt at the start actually fuels the ultimate drive to succeed. If you learn only one thing from this book, learn that when you're panicked and depressed and afraid and nothing is working right, it's okay, you're right on schedule; you're at the beginning of a potential success story.

This transformation of negative emotion into positive energy is not automatic and may not happen at all. The study showed that a successful transformation occurs in three clearly defined phases and depends on specific roles being played during each phase. This three-phase structure is visible in the generic imprint story that was uncovered during the study and also told as Jeff's childhood story in the prologue. The bare bones of this process, the most basic template of American quality, are discussed in the three sections that follow.

PHASE I: CRISIS AND FAILURE

The archetype story and its associated transformation begin in a state of crisis, what we simply characterize as a "Mess," shown in Figure 2.4. For example, at work, you are often faced with a Mess when you receive an assignment, and the only thing clear about it is its urgency. Its details and parameters, the rules, are poorly defined; they may have seemed completely clear at first, but when you start getting to work on the assignment, you find that the rules no longer make sense. This extraordinarily common situation introduces the first role in the process: the Lawgiver.

The Lawgiver is the person, group, or phenomenon that creates the crisis, in this example the person who gave you the assignment and communicated the sense of urgency. You need to take action, you want to succeed, to do it right ... but the rules aren't clear. Figure 2.5 offers examples of Lawgivers and the conditions they create.

The Lawgiver creates the pressure, forces you to move, and insists on results. You're still unsure of the rules, but because of the pressure, the urgency, you begin anyway. You leap into action, despite your vast amount of training and counseling to plan first.

This bias can be either a source of weakness or a source of strength. By following the archetype, and transforming the mess, it becomes our greatest strength. Americans are action oriented, and

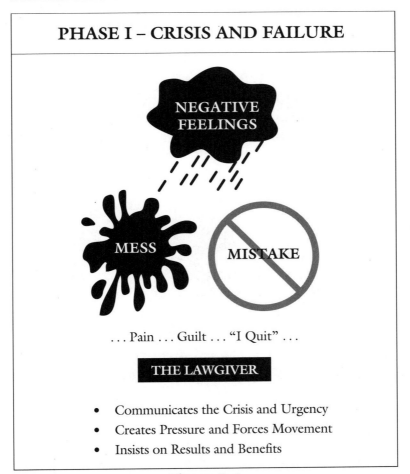

Figure 2.4

pay only lip service to planning first. They would much rather do first, plan later. For Americans, it's not Ready, Aim, Fire, but more often Ready, Fire, Aim.[1] Regardless of how much planning they do, at some point in time mistakes occur and problems result, usually from not understanding the rules or not believing them; when this happens, Americans feel bad.

This sense of failure begins the transformation—it is a critical point and cannot be avoided. It can happen relatively early or late in

---

[1]Peters and Waterman, Jr., *Search of Excellence,* p. 119.

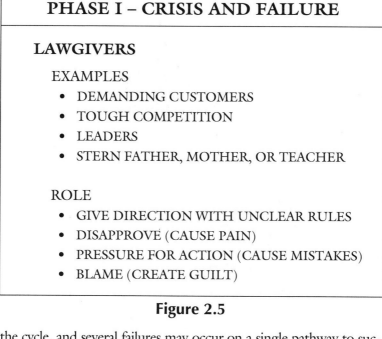

PHASE I – CRISIS AND FAILURE

LAWGIVERS

EXAMPLES
- DEMANDING CUSTOMERS
- TOUGH COMPETITION
- LEADERS
- STERN FATHER, MOTHER, OR TEACHER

ROLE
- GIVE DIRECTION WITH UNCLEAR RULES
- DISAPPROVE (CAUSE PAIN)
- PRESSURE FOR ACTION (CAUSE MISTAKES)
- BLAME (CREATE GUILT)

Figure 2.5

the cycle, and several failures may occur on a single pathway to success. The natural inclination is to try to avoid failure at all costs, but within failure lies the source of energy needed to fuel the transformation. This source of energy is emotion, particularly the emotional response to failing.

This time of emotional pain is as delicate as it is crucial. People must begin the cycle of transformation to bring them out of their initial pain and pull them toward success. If the pain is too great, they can fall into other cycles that are tragic and debilitating, preserving the pain instead of transforming it. People often wear masks, deceiving themselves and everyone around them, and may not even be consciously aware of their pain; they will almost never show it in any obvious fashion. Looking at the emotional cycles in Figure 2.6, you can see that the most critical point occurs just after experiencing failure and being overcome with pain. Everyone's first impulse is to cover up the failure and hide, but this is no escape. Even with an effective cover-up, the pain associated with failing lies somewhere inside them.

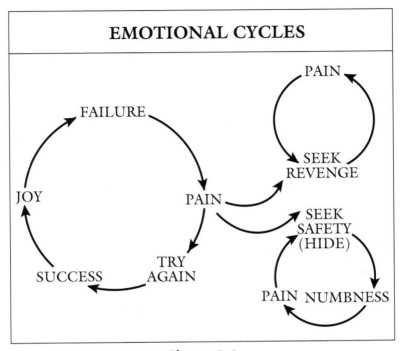

Figure 2.6

When people feel pain and move out of the healthy cycle, pictured on the left in Figure 2.6, they often move into another cycle to seek safety, sometimes reaching a state of numbness. They want to deaden the pain, diminish its impact in an attempt to heal themselves. When they find a safe place to hide, it can become home for quite some time; sticking their necks out to try again is too risky. Once they're stuck in this cycle, they may recover slowly, save face, and find a way to regain dignity, but that's about all.

Another common reaction to the pain of failing is to seek revenge. People move out of the healthy cycle into one in which the pain is masked with some sort of disguise. Here, people scheme for retribution to get even with whomever they believe it was that caused their pain, waiting for the right moment for their revenge.

All people have experienced times when they've escaped to one of these alternate cycles. Some are more prone to seek one cycle over the other, but both are very destructive, to themselves and their organizations. These alternate cycles waste time, energy, and money. But worst of all, they waste people's talent. When people escape to these other cycles, they become noncontributing members of their organizations. It's not just a matter of giving less than their best—they're giving their worst. Their potential as employees and as human beings is lost.

Perhaps the most important aspect of Phase I, the Crisis, is this: during a state of emotional crisis, people are closed to learning. They are in pain and don't want to listen to anyone. Even the most precious advice from the wisest person around will roll right off someone in the midst of Phase I. The best thing to do here is simply become aware of what a person is feeling and face it honestly.

PHASE II: SUPPORT

As Phase I is characterized by a Mess, Phase II is characterized by Support. In Phase I, you've been thrown into a crisis because the rules from the Lawgiver were not clear. You burst into action, not asking for a plan or clarification. You were well intentioned and wanted to do it right. You took risks and gave an honest effort, but it didn't work. You failed, and by now you're probably feeling a certain amount of guilt. This phase, Phase II—Support, introduces the second important role in the cycle of transformation: the Mentor, described with some key characteristics in Figure 2.7. (See Figure 2.8 for examples of Mentors and the roles they play in the support phase.)

Listening to their problems, and acknowledging the way they feel about them, the Mentor helps people accept the reality of their situation, and in so doing helps them release any guilt they may feel. In this way, the Mentor helps them renew their sense of purpose. With the Mentor's help, people become willing to try again and again, as often as necessary, to continue along this path, until they achieve success. Cycling through failure several times is certainly not uncommon, but the Mentor must be there every time to offer the necessary emotional support.

In Phase II, the Mentor is accompanied by another very important role, the Coach. While they may sound like the same role, the

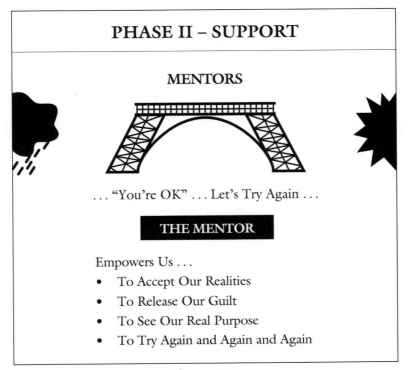

Figure 2.7

study showed that Coach and Mentor are noticeably different roles, even though they're occasionally performed by the same person. Note that while one person can act as both Mentor and Coach, it is inadvisable for a Coach or Mentor to also be a Lawgiver. It is recommended that the Lawgiver be some outside force, like your competition or your customer, although a business leader may need to become a Lawgiver to ensure the competitive threat or that the voice of the customer is heard before it is too late.

The role of the archetype Coach corresponds to the coach of an athletic team, an analogy we follow in Figures 2.9 and 2.10. The Coach works with people to break projects down into smaller, more manageable steps and helps them develop their skills and acquire whatever resources they need to do the job. The Coach helps people achieve small, measurable successes, one at a time, build on them, and eventually achieve the larger success of the whole project. The Coach

---

PHASE II – SUPPORT

MENTORS

EXAMPLES

- GRANDFATHERS AND GRANDMOTHERS
- OLDER SISTERS AND BROTHERS
- CARING PARENTS, TEACHERS, OR COLLEAGUES
- COWORKERS
- COACHES

ROLE

- RELIEVE GUILT (NOT YOUR FAULT)
- HELP FIND SELF AND PURPOSE
- PROVIDE MEANING AND VALUE
- SERVE AS EMOTIONAL CATALYST

Figure 2.8

also sets deadlines and insists that people practice. When was the last time your business team practiced or trained together as a team?

In business, people use sports analogies all the time. They look for examples in football or baseball that are clear and obvious, and then translate them into the language of the business world. But there's one glaring difference in the sports versus business analogy, a difference managers never talk about when encouraging their staff to go-team-go. Of the total time an athletic team spends together on the field, 80% to 90% of it is spent in practice, learning how to do it right by doing it wrong the first time—in practice. Only 10% to 20% of their time is spent actually doing it right—competing. In contrast, business teams spend nearly all their time trying to compete in the real world, with real customers and real money, and spend only 2% to 3% of their time practicing, if in fact they ever practice. Business teams compete far better when they get a reasonable amount of practice; this is why they need coaches.

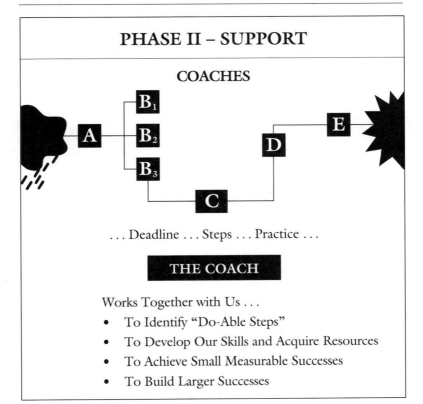

Figure 2.9

All executives, managers, and supervisors should practice some degree of coaching. In today's business environment, many people believe that good coaching is only found outside the workplace, usually at some type of special off-site training session. A manager may give an employee a little impromptu coaching during a performance review, once or twice a year, but on average that's about it. It's far more advisable for managers to provide regular, frequent coaching in the workplace, where people can get it when they need it.

In the course of Phase II, the Mentor provides emotional recognition, which relieves people's pain, helps them release their guilt, and opens them to learning. Then the Coach steps in to teach them, deepen their understanding, provide the right tools for the job, keep them on schedule, and make them experiment and practice

---

PHASE II – SUPPORT

COACHES

EXAMPLES
- BOSSES AND COLLEAGUES
- PARENTS, BROTHERS, AND SISTERS
- TEACHERS
- MENTORS

ROLE
- BUILD SKILLS
- INSIST ON PRACTICE
- PROVIDE SUPPORT
- ROLE-MODEL BEHAVIOR
- ENCOURAGE LEARNING FROM EXPERIENCE

Figure 2.10

safely, away from the customer. This lets people refine their plans and methods to do the job right eventually, for the customer.

PHASE III: CELEBRATION

When Americans persevere long enough, and receive the right kind of support from Coaches and Mentors, the transformation ends in success. This third phase of the archetype is a Celebration of the quality performance, shown with supporting details in Figure 2.11. Phase III introduces the last key role in the archetype structure: the Champion. The Champion is you, or me, or us, the workers or players, the ones who have been struggling to beat the odds, to make the deadline, to win the contract, to finish the job, to ultimately overcome the pain of initial failure and finally succeed. Figure 2.12 lists the Champion's most important characteristics.

PHASE III – CELEBRATION

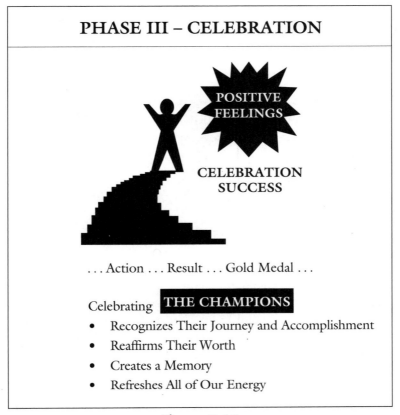

... Action ... Result ... Gold Medal ...

Celebrating **THE CHAMPIONS**

- Recognizes Their Journey and Accomplishment
- Reaffirms Their Worth
- Creates a Memory
- Refreshes All of Our Energy

Figure 2.11

The first thing to understand about a Celebration is that it's not a party, but it may have a party feeling or atmosphere. It is more like a ceremony and sometimes acquires the flavor of ritual. It's a time to acknowledge the Champion or Champions who have achieved their success. The ceremony must recognize not merely the Champions' wonderful accomplishment at the end but their entire journey, including all the failures, all the pain, the desire to quit, and the struggle to persevere. When people recognize the whole journey, they acknowledge that this experience is human. Parents, teachers, managers, and executives must show their Champions that they accept the Champions' initial failure, accelerating progress, and ultimate success as a single, nurturing process,

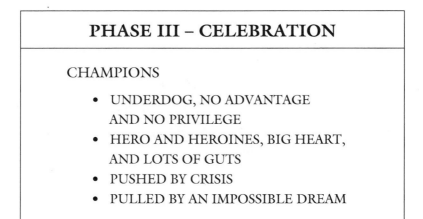

Figure 2.12

not discrete events to be analyzed individually. By reliving their journey through storytelling, and sharing their feelings, the Champions prepare for their next challenge. This is essential!

The Celebration should be planned so that all the people attending can identify with the Champions' story. Remember, the Champions weren't always winning. Sometimes they were getting knocked down, sometimes stumbling and falling, sometimes making mistakes and feeling horrible. During the Celebration, those in the audience relive this journey with the Champions. Even though the story doesn't focus on their own experience, they will still identify with it. When the last part of the story, the successful part, is the only story told, the Champion is set up as superhuman. Those in the audience feel separated from the Champion, rather than seeing that they also could be in the spotlight.

In all likelihood, the people in the audience are cycling somewhere in journeys of their own, perhaps still in pain, perhaps in failure, maybe wanting to give up. Attending the Celebration of another group or individual can provide them with tremendous emotional energy. Those seeking to be effective Mentors should be aware of Celebrations taking place in their vicinity (social or corporate), in the event that their own people might benefit from this sort of lift. Sometimes a vicarious Celebration is just what people need to pull out of Phase I and begin again.

At this point, it's worth repeating that the Celebration we're talking about is not a party. If you benefit from attending someone else's Celebration, it won't be due to free food and a chance to leave the office a few hours early; what you're after is emotional energy. A Celebration reaffirms the worth of the Champions, and a subliminal echo of this acknowledgment tends to spread to all the people in attendance. The Celebration makes it clear to the audience that the Champions are considered people of outstanding value, people who are respected for having lived through the whole transformation, and that means respected even for their initial failures: at least they tried.

These Celebrations are extraordinarily emotional. When the Champions tell their stories, including all their emotional highs and lows, they create a powerful emotional memory which reinforces the imprint within the archetype. Powerful emotions lead to a powerful and compelling Celebration, and the strongest emotions create a very deep imprint of the experience. It refreshes and stimulates the Champions' energy, and the energy of everyone in the audience.

Through the Celebration of Phase III, the Champions internalize their learning from Phase I and Phase II. The Celebration reinforces the structure of the archetype and releases the energy needed for the Champions to take on new challenges. Over time, this sort of recognition can accomplish the priceless task of reducing people's fear of failure and their immobilizing reluctance to take risks.

Figure 2.13 shows the complete transformation during the three phases and the learning that occurs during each phase. Notice that the negative feelings in Phase I are balanced by the positive feelings in Phase III. This is not merely a graphic designer's trick, it's what actually happens. Americans must achieve this balance. Each of these elements is part of the American quality archetype. Each element must be present for any American quality initiative to be successful. It's a simple model and a powerful nurturing process, one that offers enormous potential.

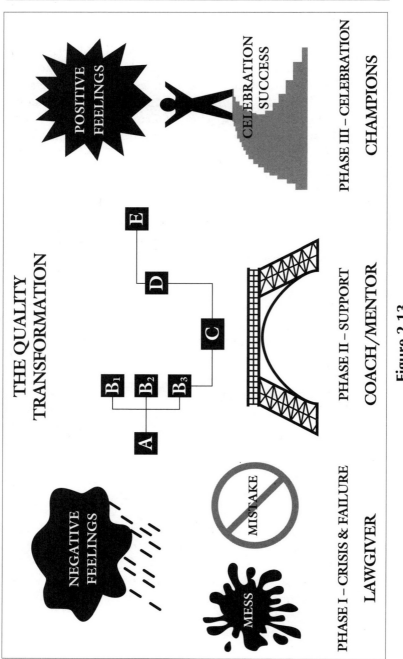

Figure 2.13

AMERICANS AND PERFECTION

We can demonstrate another critical element of the American quality archetype by relating typical American behavior to the "Q" line shown in Figure 2.14. The Q line is a simple graphic depiction representing marketplace expectations. It moves up because customer expectations of quality, as set or demanded by the marketplace, are always moving to a higher level. As new products, services, technologies, or levels of quality are achieved, these expectations continue to rise. This line therefore symbolizes quality as constantly moving and evolving. The study showed that for Americans, the quality of goods and services is most often associated with the phrase "It works," meaning that a product or service does what it's supposed to in the way people expect.

Contrast this American idea of simple functionality with perfection. You've read what perfection means to Americans: the end, death. But for the Japanese, the Q line appears to be more closely aligned with true perfection—the results of this difference are quite

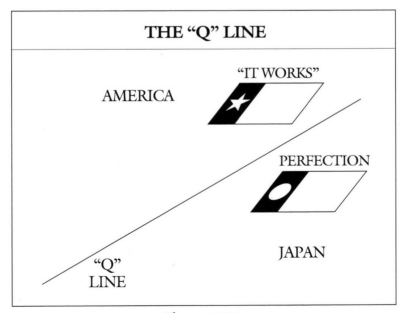

Figure 2.14

striking and have a remarkable effect on their performance as producers.

In both the United States and Japan, the farther you are from the Q line, the farther you are from meeting marketplace expectations; your processes, products, and services are full of defects and deficiencies. This means you're in a greater state of crisis and have a greater awareness of the need for urgency. The big difference between Japan and the United States at this critical stage is the length of time before the reality of the situation is fully acknowledged. It often takes death or imminent disaster to shake Americans awake, which leads to tragedies like the *Challenger* explosion, the Exxon *Valdez* oil spill, or Three-Mile Island. The Japanese appear to be more open to feedback and less likely to cover up problems and hide from criticism.

When awakened to a crisis, whether sooner or later, both Japanese and Americans rapidly take action to approach the Q line. Because of cultural differences, the Japanese can and will continue to work with great diligence, dedication, and constancy of purpose to get closer and closer to this marketplace expectation, as shown in Figure 2.15. The consistent discipline of the Japanese enables them

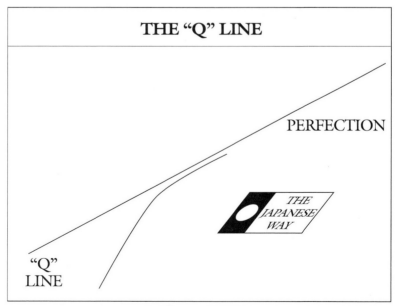

Figure 2.15

to squeeze out every last bug, problem, and deviation ultimately to achieve close to zero defects. In fact, they are very likely to push past the point of exceeding current market expectations, thereby setting new thresholds.

For the Japanese, constantly approaching the Q line is the equivalent of striving for perfection, in their culture a wonderful source of positive energy. When people in Japan achieve excellence or set a new standard, they become highly respected and are often revered as masters. Such people achieve a position of the highest regard in Japan and receive tremendous respect, status, and recognition. As a reward, masters are permitted to deepen their wisdom by pursuing their work for the rest of their lives and are expected to teach other people their priceless skills and knowledge.

In this country, it's quite different. Once Americans achieve the highest level of excellence, they certainly don't want to continue doing "it" for the rest of their lives; they get hopelessly bored. For Americans, getting closer and closer to the Q line is anything but a source of positive energy. Americans don't like being methodical. There is little recognition for the diligence and perseverance required to maintain quality in the same task or product. In Figure 2.16, the dotted line shows what happens to Americans when they try. Instead of getting closer and closer to the Q line, boredom sets in, defects multiply, and the curve falls off.

In America, if people are losing enough market share, if they are in pain and crisis and are willing to admit it, then they will move. They'll feel the urgency and the crisis, and they'll begin trying to reach the Q line. But once again, as they get closer and the end is in sight, they lose interest. Why? Because to reach the line and maintain a position there requires precision, control, discipline, and squeezing out every last defect, an ongoing task that holds little excitement or reward. When Americans get close to the line, they're likely to say, "Well, looks like we're almost there. We can make it." They relax and begin looking for something new, something more exciting. Itching to move on to a new challenge, Americans get sloppy. Their interest and discipline wane, and they start to fall.

As you might guess, this finding caused us great concern. Fortunately, it surfaced early in the study, allowing us to explore it in greater depth. During this exploration, we discovered the last

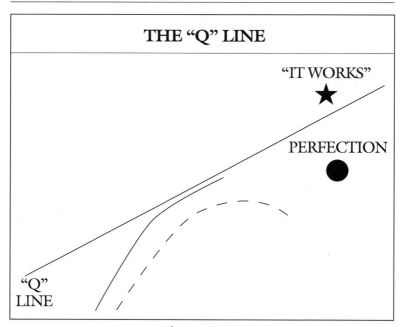

Figure 2.16

element of the archetype, something that turned out to be characteristically American and perhaps should have been obvious from the start.

Figure 2.17 illustrates the way quality works in America—quality the American way. It begins with a great push from the crisis position, the low point full of defects and mistakes. With a feeling of urgency, Americans begin to move forward. What's needed to pull them up to the Q line is actually something that lies well beyond it: an impossible dream. This can't be just any dream thrown at people like some emotional panacea. The right impossible dream for Americans must bear certain characteristics.

First, it must be something that seems almost literally impossible to achieve but is eminently worth striving for, like an Olympic gold medal. Second, it must inspire that itchy, tantalizing feeling that despite its loftiness, maybe, just maybe, people can do it. A dream can't be too far-fetched.

People's personal dreams are easily connected to this greater impossible dream which can appeal to many. It's the challenge that

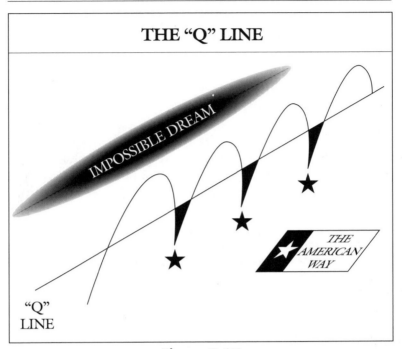

Figure 2.17

excites people, and it must be a significant and distinctive challenge, preferably something unique. Meeting the specification that already exists is boring; doing the impossible, that's another story. But for Americans, the energy from such a dream has its limits and its pitfalls.

Even with the addition of an impossible dream, Figure 2.17 shows basically the same pattern as Figure 2.16. As Americans get closer and closer to the dream, maybe even reaching it, they lose interest. They see they're almost there, so they get overconfident and sloppy and become more interested in finding something new, something more challenging. With this overconfidence, a lack of precision and discipline creeps into their work, and it falls away from quality, sometimes disastrously. How far below the Q line the work will fall is critical. How much market share did American auto makers have to lose to the Japanese before they took action and tried to implement change? How much pain do Americans have to feel before they're willing to admit that they're failing, that they're in a state of crisis? The archetype study showed that a crisis creates negative, painful emotions, which can be redirected into the positive energy needed for tackling the next challenge. But what most Americans tend to do instead is dissipate the negative energy by calming the crisis, searching for the smallest crumb of good news buried in the bad to make us feel better.

The vital step here is to create a sense of urgency before performance falls below the Q line. Channel this negative emotional energy by connecting it to a new impossible dream. This can rekindle people's spirits. In such an effort, the crisis ignites people's emotional fuel, driving them forward. Mentors help them connect with the new and exciting purpose, and the impossible dream pulls them on to win.

Today, as American parents, teachers, managers, executives, and officials struggle for success in home, education, business, and government, they move from peak to valley, from highs to lows, lows to highs. It's a persistent roller coaster, and it has to be, at least to some extent. If Americans are to maintain consistent improvement in products and services, this variation is absolutely essential. Whenever those in charge of a project try to flatten the emotional curves, believing that this is what is needed to maintain some sort of "controlled quality," people lose their excitement, their energy, and, finally, their interest. This is exactly the way to destroy the value of the greatest treasure in the American workplace: Americans.

The American space program is an excellent illustration of this quality process. When the Russians launched *Sputnik* in 1957, they set a new standard for achievements in aerospace. The Q line for that particular arena moved up, and Americans were no longer the leaders—the Russians were. The gap that existed between the two countries was enormous. America was the underdog, very much behind; that was very painful for us. But actually, this is the best position in which Americans can find themselves. They don't like being second one bit and don't want to stay there any longer than necessary.

In 1960 and right on cue, President Kennedy stepped in with an impossible dream: putting a man on the moon and returning him safely to earth before the end of the decade. With Americans' drive fueled by the pain and embarrassment of being the underdog, and their will and determination inspired by Kennedy's vision, they set out on a mission to be the first to conquer space, and they did! America left *Sputnik* near the starting line, landed a man on the moon, and returned him safely to earth in July 1969.

Sadly, even with these glorious circumstances, space exploration followed the pattern of the archetype. Sloppiness crept in, and the precision and discipline were gradually lost. The program's quality slipped, then fell. First, a fuel cell explosion crippled the *Apollo 13* spacecraft in April 1970. The three astronauts barely managed to improvise control and environmental systems to stay alive and return to earth. In 1979, due to an inaccurate long-term mission plan, *Skylab's* orbit began to decay earlier than expected. As the spacecraft began to break up, the whole world wondered anxiously where the possibly large chunks of metal would fall. Through nothing but luck, no damage was done.

But that still wasn't enough to fix the space program; it continued to sink below the Q line. Urgency and crisis were felt, but the need to stay on schedule was the primary driver, and more people had to die before NASA would admit a state of crisis. It took the *Challenger* tragedy to call everything to a stop. Today, the space program continues to flounder with one failure after another. Why? Because it has its crisis, but no dream.

QUALITY ON THE OUTSIDE

Once upon a time, a young man was helping his father, who was in the construction business, with a particularly difficult roofing job. After they had finished, the father spent some time reviewing the work with the customer, and then told his son they had to tear it apart and start over. The young man was livid; he and his father had worked very hard and done a good job. Why should the customer's opinion be their problem? The son's reaction was, "That's too bad, they'll just have to accept it." When the son continued to protest, his dad said, "Listen, son, the lesson here is: sometimes in order to win, you have to lose. Let's get back to work." The story has a happy ending. The customer was so delighted with their work that referrals came pouring in to the father's business, and the son learned a novel and powerful lesson from his father. Many people preach a win-win strategy, but perhaps a lose-win strategy is the best way to satisfy a client to achieve a win-win in the long term. It all depends on how much you care about your customers.

Because the quality archetype for Americans includes failure and the pain connected with that failure, products and services do not have to be defect free for Americans to judge them to be of high quality. The archetype study showed that, with the right circumstances, Americans perceive even products and services that are not defect free as possessing high quality, if they feel they are cared about by the producer when a problem occurs. This is a powerful insight. People who care, really care, for their customers can build loyalty for life. Americans want to feel that producers care about them. They want producers to feel concern when they are caused trouble and inconvenience due to defects in the producer's product or service. This situation applies whether the customers are internal or external to an organization. Note that while caring is critical, people have a limited tolerance for reoccurrences of a problem.

A product or service provided to customers is symbolized in Figure 2.18. As the producer of this product or service, you see it through quality tunnel vision. It's a first-rate offering, defined by clear, fixed groups of specifications, features, and benefits, full of wonderful advantages for the customer. Customers, however, intuitively know this offering is not perfect. Customers have reason to feel somewhat uneasy about the item before they even buy it.

Too often, producers forget that the boundary around a product or service extends beyond its physical, quantifiable dimensions. This boundary encompasses something you cannot see or touch but which

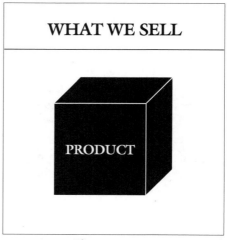

Figure 2.18

customers experience through intuition: anxiety. Even if a consumer's past experience has not left its mark of suspicion, the mental structure of the archetype has left an imprint. From the moment Americans become active in the marketplace, usually as teenagers, they accept without even thinking that any product or service they obtain will not be perfect. The anxiety stems from their concern over what's going to happen when these hidden imperfections appear. Will their rented car merely hesitate a little when accelerating from a full stop, or will it die suddenly without warning, stranding them on a lonely stretch of highway? Will the snapshots of their daughter's wedding merely be poorly printed or will the negatives be lost at the photo lab? This relationship of product and associated anxiety is shown in Figure 2.19.

The focus of modern business should address the whole package—the product or service and the anxiety that surrounds it. Companies that deal honestly with this anxiety win long-term loyalty from customers. They demonstrate their concern before and after a product or service is sold and especially when problems appear. Companies that demonstrate care and concern are those to whom customers return over and over again, even for entire life-

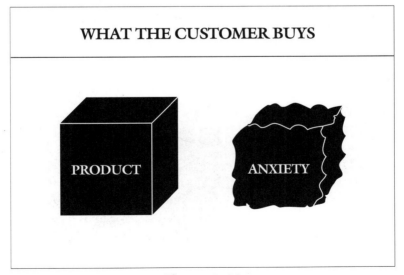

Figure 2.19

times. Sears, with its lifetime guarantee (no questions asked) on Craftsman tools, wins allegiance and loyalty because they show concern for their customers as people. They accept and honor a certain degree of responsibility, at least as far as their products or services permeate their customers' lives. In short, they care.

HIGH-TECH PANIC

High-technology companies, makers of everything from computers and jet engines to consumer electronics, have developed their own perspective on the issue of customer anxiety. They see their product design and associated customer support as dependent on an inverse relationship between high-tech and high-touch (high-tech indicates a level of technical sophistication; high-touch, a degree of customer hand-holding by the producer). The dominant belief about these qualities and their relationship is shown in Figure 2.20. Basically, this belief states that the more technology you build into a product, the more intelligent it is, and therefore the less you need the human intervention of customer hand-holding. For example, a super-sophisticated VCR, replete with microprocessors and automated features, should need virtually no human support after the sale. Applying the knowledge from the archetype study, you can see this assumption is dead wrong: smarter products don't require fewer support people, they require more.

The flaw in this widely accepted belief lies in the assumption that the more intelli-

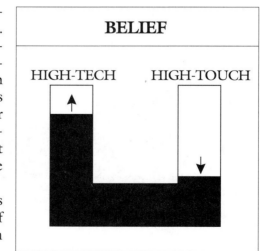

Figure 2.20

gent a product is, the easier it is for customers to use; most often, this is not the case. An industrious and clever young man in New York City recently turned this very problem into profit. He created an overnight business success by providing house calls to busy New Yorkers to show them personally how to operate their sophisticated, high-tech VCRs.

Considering that customers buy a layer of anxiety with every product or service, it's a mistake to think that smarter products produce less anxiety. Smarter products offer more, do more, are capable of more, and therefore have many more things that can go wrong. This makes customer anxiety about that fancy new VCR far greater than it would be for, let's say, a toothbrush. A related issue is the problem of smart products making customers feel stupid. Products that reasonably intelligent customers cannot operate properly, even after hours of pressing buttons, flipping switches, and reading the owner's manual, create grim clouds of anxiety. When these same products break down within a few weeks of purchase (or worse, a few weeks after the warranty has expired) and no service is provided, the company has lost a customer for good. This more realistic relationship between high-tech and high-touch is depicted in Figure 2.21. The more technology a product offers, the more human contact and support the customers need. It would also be better for designers and manufacturers to start thinking of intelligent products as those that make their users feel intelligent.

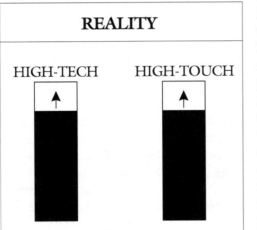

Figure 2.21

LOYALTY

Over and over, research shows that customer loyalty, the kind that leads to enduring, long-term relationships, is built on Phase I and Phase II experiences: failure followed by support. Just as the crisis of Phase I in your own life is a time for you to channel your emotional energy, problems that customers experience are wonderful opportunities for you to demonstrate how much you care about them, thereby winning their loyalty.

Businesses invest heavily in programs to handle customer complaints. In high-tech companies, such tiered customer support includes elaborate procedures for handling equipment failures. Such firms maintain highly paid, technically skilled teams, whose trouble-shooting and problem-solving talents are valued for their competitive advantage. Given what the archetype has taught us about customers and our own learning process, we see a unique opportunity here. With a simple addition to existing customer service routines, companies can achieve increases in customer loyalty and refinement in their handling of equipment failures in a single operation. All that's required is something that alleviates our customers' high-technology anxiety, while taking advantage of the natural American tendency to learn by doing. Consider the following idea.

When you deliver a product or service, plan a breakdown rehearsal shortly after customer acceptance. This rehearsal should be a live, realistic simulation of whatever the customer's service agreement calls for, in the event of a serious failure. When you call the customer to schedule the rehearsal, assure them that there are no problems and you anticipate none, but just in case a problem appears, you'd like them to know exactly what will happen so they can sleep peacefully at night. This effort, in addition to stunning most customers into silent disbelief, will prove to them more than any sort of lip service that they are indeed in the hands of caring, sensitive people, people who are going to remarkable lengths to assure quick resolution should any problems arise. When you execute the rehearsal, stage it as a full simulation, down to the last phone call, power outage, and replacement part. This will not only provide unparalleled good feelings for your customer but will also let you find the flaws in your procedures, building sure-footedness and confidence among your customer support people.

Figures 2.22 and 2.23 each show similar results from two different studies. The message in these data is that effective handling of customers' problems does not degrade your relationship. Quite the contrary, it solidifies and enriches the relationship. Customers are far more loyal to companies that have handled their problems well than to companies where no problems have ever occurred or that deny that it is their problem when the product fails to work properly. As mentioned earlier, this behavior is a manifestation of the archetype. Unconsciously, customers do not expect or even want perfection—they want things that work the way they're supposed to, and when they don't, they need support from the provider.

Donald E. Petersen, retired chairman of Ford Motor Company, has said, "I am convinced that the passion of the employees begets the passion of the customer." He noted the vitally important link between technology and emotion, citing the two aspects of customer emotion with which all companies must concern themselves. On one side are the positive emotions—delight, contentment, or security—that products and services should inspire. On the other side are anxiety, apprehension, or other negative emotions that customers feel until they experience proof of something's quality. To

REASONS CUSTOMERS LEFT COMPANY

70% HAD NOTHING TO DO WITH THE PRODUCT

- 20% SWITCHED BECAUSE THEY HAD TOO LITTLE CONTACT AND PERSONAL ATTENTION
- 50% SWITCHED BECAUSE THE ATTENTION RECEIVED WAS RUDE AND UNHELPFUL

Source: The Forum Corporation

Figure 2.22

Figure 2.23

eliminate the customers' anxiety and assure their delight, he said, "The company must generate a positive emotional force from within ... I believe a company can, to its benefit, capture the employee's emotional involvement."[2]

In this way, we cover all bases. Acknowledging the emotions of employees and customers, and providing mentoring and coaching as needed, can provide the fuel that can drive America's success in the marketplace. An acceptance of these emotional truths, and an honest application of the archetype, will gain equal loyalty on both sides of the business equation: our own people will give their greatest efforts and most visionary ideas, and our customers will bring us their business again and again.

QUALITY THROUGH THE AGES

There's a certain pattern that typifies the American pursuit of quality. It's a pattern that often applies as much to our individual efforts as it does to the entire history of quality improvement.

---

[2]Donald E. Petersen, from a speech delivered at the Conference Board's Marketing Conference, New York, November 1988.

Since the Industrial Revolution, the main thing that quality-minded Americans sought was technology: steam power instead of wind or water, electricity instead of gas, and plastic instead of wood or metal. People believed they could do a first-rate job, and produce high-quality products and services, if they just learned and applied the right technology. Even today, the dominant attitude among producers and customers alike is a hunger for new things that are visibly products of new science (take a look at the latest Sharper Image catalog). In business, many are the novice managers who insist on new computers, space-age materials, or elaborate retooling as the best way to attack a quality problem.

More recently, Americans began to see that technology alone was not enough and began to add new elements to the quality equation. They realized that they needed good planning, people to write the plans, and others to turn the plans into action. Again, the growth of the individual manager in today's business world often parallels this societal evolution.

After implementing this approach, it became clear that this still wasn't enough. American companies began to feel the nibbling of foreign competition, and most realized that plans and technology alone could not both achieve and maintain satisfactory levels of quality. Companies began to feel that they needed leadership as well as management. They believed that when leaders became involved, offered vision, walked their talk, set the direction, and became truly committed, then they'd surely be able to move forward with powerful quality initiatives that produced results in the marketplace.

This was a big step. The business world acknowledged the work of many fine leaders, some of whom wrote extensively about the importance of enlightened business leadership. This was very close to the complete picture.

All along, the missing element has been an acceptance of the importance of human emotions. Consider emotions as a symbolic representation for people; somehow, just saying people doesn't convey the right idea. This is a tough concept for many people to accept. In business, Americans are perfectly comfortable with intellect and have been for generations. They must become equally comfortable with emotions.

The traditional ranking of these quality improvement approaches is as follows:

1. Technology (great!)
2. Plans (even better!)
3. Leadership (now we can't lose!)
4. Emotions (huh?)

The archetype research has shown that a more pragmatic and demonstrably more effective ranking is as follows:

1. Leadership (great!)
2. Emotions (of course!)
3. Plans (now we know what to do!)
4. Technology (let's do it!)

The American business environments in which quality flourishes are those in which the leaders inspire their people, capturing their hearts, souls, and emotions. The people then channel their emotional energy to power up their plans, approaching the marketplace realistically but urgently. Finally, the people execute their plans with the right technology, creating products and services to win and delight customers. American approaches to quality must be balanced in this fashion. The emotional dimension must be involved at the beginning, from the top down, and all behaviors and expectations of one another must be aligned with the archetype.

Take away my people, but leave my factories, and soon grass will grow on the factory floors. Take away my factories, but leave my people, and soon we will have a new and better factory.

—Andrew Carnegie

— ✧ —

Coming together is the beginning. Keeping together is progress. Working together is success.

—Henry Ford

— ✧ —

Real courage is when you know you're licked before you begin, but you begin anyway and see it through no matter what.

—Harper Lee

3 LEW'S PARALLEL UNIVERSE

TROUBLE IN ATLANTA

In April 1984, while Marilyn was in New Jersey, soon to begin her frenetic struggle to find a new job in the fledgling Network Systems quality organization, I was down in Atlanta managing the Copper Shop, the old Western Electric cable production facility. At this point, it was nearly two years before Marilyn and I would meet, and my attention was firmly rooted in the present. At the Atlanta Works, we were going through a struggle of our own.

As a result of AT&T's divestiture that same year, major changes began to take place in Atlanta, and our business unit felt the impact immediately. It wasn't long before upper management significantly reassigned our copper-exchange cable product lines; these lines encompassed special cables designed for telephone exchange sites, the so-called "Local Offices" that accomplished the first level of call switching for individual customers. Manufacture of these products would be phased out of AT&T factories in Omaha and Baltimore and consolidated at just two locations: the Atlanta Works in Georgia and the Phoenix Works in Arizona. Besides copper-exchange cable, the Atlanta factory also manufactured the newest telecommunication products, including fiber optic cable and related apparatus.

That year, we had to work at a feverish pace to install transferred production machinery from the other factories; this was an essential operation and also seemed to be to our advantage. It would allow us to expand our manufacturing capacity coincident with the phaseout of cable production at the other factories.

We were putting in a lot of overtime but still couldn't keep up with the copper cable demand. Adding people helped but resulted in so much movement of personnel that a massive training effort had to be mounted at the same time. To make an exceedingly long story short, by year end the copper cable orders dropped to what we considered a normal seasonal level. We all felt relieved. This was the break we needed, giving us a chance to catch our breath and get ready for the usual spring rush of orders.

In the first few weeks of 1985, our Copper Shop was riding high. We had installed the new manufacturing facilities and had successfully taken on all the production phased out of the two other sites. The speed with which we completed this change gave us a tremendous amount of satisfaction. There were a few minor quality problems, but we knew that time and training would smooth everything out.

A WRENCH IN THE WORKS

Our elation was short-lived. In February 1985, our seven-member management team became aware of a potentially disastrous trend: due to a combination of the new postdivestiture competition, and the rapidly increasing popularity of fiber optic communications, copper cable business was on the decline. Only months earlier, we were frantically adding machines to increase our capacity, and now they were sitting idle. By the end of March, the critical nature of the situation became evident. Business was dropping, and we felt an increasing urgency to take some sort of action.

Nothing was going right. What were originally minor quality problems escalated into major ones. In April, the independent AT&T quality assurance organization rated several of our product lines "out of control" and put the operation on "quality alert." This triggered a series of procedural and administrative systems that created the equivalent of "Big Brother," watching over us constantly, and there was nothing we could do about it. Until the situation was completely rectified, top management outside the Atlanta Works would be asking lots of questions and looking for immediate corrective action.

Conditions became worse. Lower orders meant we didn't need as many production specialists; these were the people who made up most of the Copper Shop's rank and file. In an attempt to balance this situation, we declared a surplus condition in copper operations and transferred excess workers to the fiber optic shop. An overlooked and problematic side effect of this decision was the loss of experienced copper cable employees. This made the whole copper operation less efficient and damaged our already dreary financial status. Our poor performance became a major issue, and in May, the corporate planners met with me and painted a very grim picture indeed. It seemed that our sister location in Phoenix was producing better financial results. If copper business continued to decline, the Atlanta Copper Shop would gradually be phased out and eventually closed. Because I believed we could do little except wait to see what would happen among our customers, I chose to keep this information to myself. If the need for copper cable began to increase, we were saved. If it continued a steady decline, we were history.

By the end of May, we had settled into a more or less stable routine of running the shop, watching our performance and hoping for more orders. Things didn't improve, but at least there was a certain day-to-day continuity.

Then, with no warning at all, the planning organization decided to expand our fiber optic operation. While on the surface this might seem like good news, such executive decisions tend to get messy down at the implementation level; this was no exception. To accommodate the new fiber optic machines, we would have to pull out some vital copper insulating machines to provide enough space—another crisis.

My experience during these months at the Copper Shop was full of excess stimulation, and in time this began to have some noticeable effects on me. It made me think of what battlefield combat might be like and how relationships and personal commitments form when bullets are flying over people's heads. The events of the first half of 1985 were anything but business as usual for the Copper Shop. We all knew it would take some radical solutions to get us back in the black. One thing that happened was that I began working much more closely with my boss and his boss, Walt Ehmer and Lee McClary, respectively. This turned out to be a valuable if not essential trigger for the events that followed.

THE BEGINNING OF SUPPORT

A great benefit of working very closely with people under conditions of stress is that you get to know them better—you can never be too close to colleagues. I had known them for years, but during these months we crossed some sort of emotional threshold. We became even closer, which turned out to be a big help for all of us. During this period, I learned with some surprise that they were both just as concerned about the shop as I was, so I no longer felt so alone in my worries.

As our relationship evolved, the level of communication we shared changed a great deal. A year ago, we had shared a strictly boss-subordinate relationship and had talked only about the job. But during the last few months, we had become far more open with each other about some very sensitive issues, things that we would have been reluctant to discuss a few months earlier. We began to talk more like friends than coworkers and freely discussed topics concerning our families and personal lives. This made it much easier to express feelings and ideas and went a long way toward removing emotional barriers to communication. There were no unsafe topics—we could all say anything about anything, no matter how critical or potentially inflammatory.

Walt and I admitted to each other a deep, sad frustration with the ongoing situation at the Copper Shop. Nothing we tried was working, and we tried a lot. Every day seemed to be a fire-fighting session, no matter how much we tried to smooth things out, no matter how much we planned ahead in an effort to avoid problems. We tried to take control of more and more areas and operations, believing that only by having more direct involvement in shop activity could we turn the tide and improve our performance. This seemed to have the opposite effect. The more we tried to take control, to become more directly involved in shop operations, the more we spent our entire days just putting out fires.

Without really thinking about it that much, we began building a working relationship based on accepting, appreciating, and encouraging each other to share experiences and express feelings honestly. This led to better insights into our own strengths and weaknesses, and more valuable still it created a new working atmosphere, one in which honesty was encouraged and accepted rather than discouraged and feared.

Up to this point, Walt and I had practiced the time-honored business skill of management by coercion. Under the best of circumstances, this approach has its limits, and with the Copper Shop's mounting crises it was clearly time for a change in the way we did things. We decided to expand our new-found approach to business relationships to the rest of our departments and tried to adopt a management style founded on listening and learning.

The management team began spending more time on the shop floor, on all three shifts, so we and our employees could get to know each other better. Our intentions were the very best, and my managers were giving it all they had, but we still delivered a lot less than we promised. Every day brought a new crisis, and even working 12 to 14 hours at a stretch didn't keep conditions from getting worse. Despite our stronger, closer, and more honest relationships, we were unable to sustain our energy. Gradually, we became short tempered and noticeably frayed around the edges. Worst of all, finger-pointing and turf battles became a daily occurrence. Each manager felt pressured to improve his department's efficiency, even at the expense of other departments in the operation. Our team spirit was beginning to collapse instead of grow.

My gradually expanding involvement with quality had begun in earnest in June 1985; despite all my other problems and responsibilities, I had been asked to join a team charged with developing a total quality plan for the Atlanta Works. I clearly didn't have the time but decided the smartest thing would be just to grit my teeth and do what I could. I agreed to participate, with little if any enthusiasm.

Back in April, as part of my training and orientation for this total quality project, I had been sent to a four-day quality seminar conducted by Dr. W. Edwards Deming. If you don't know Dr. Deming's reputation, suffice it to say that he's the man given credit as the single most influential person in Japan's astonishing economic turnaround and international success. The national quality award in Japan is named the "Deming Prize." His teachings are dangerously radical by some people's standards, intuitively obvious by others. What mattered to me was that he completely redefined virtually every aspect of work and the workplace, including the roles of management, the customer, and even business itself. It was fascinating and powerful learning, and I drew energy from it like a starving man. I felt like a plant that had been trying to grow under a rock suddenly exposed to sunshine.

During this seminar, I began to believe there was a better way to do things at the Copper Shop. For months, we had been going through a series of crises, had felt a lot of pain, and were still unsure of what to do to make the business survive. I often blamed the people under me, the supervisors and production specialists, for our manufacturing problems, but one of Deming's major contentions was that management created 85% of the problems. Could this be true? The possibility that I was part of the problem was a strange and not very comforting thought.

Another of Deming's key principles was the absolute necessity of eliminating fear in the workplace. Employees must feel no fear. They must not fear to ask questions, to raise issues, to seek further instructions, to mention problems or express ideas; and they must believe their ideas will be valued, not merely tolerated. The lowest-level employees must feel free to approach the highest level executive to discuss a problem or offer a new idea. It's these people who are closest, day after day, to the real work of production. They know it better than anyone and are a manager's best source of information on how things are going. But they will only speak if they feel unafraid, free of possible criticism, demotion, or imprisonment in some organizational cul-de-sac.

When I asked myself how this idea related to the Copper Shop, I didn't like the answers. Walt and I thought that walking through the shop and rubbing elbows with the production specialists and supervisors had been such a great, enlightened thing to do. Who did we think we were kidding? The people on the floor knew very well why we were there: to find something wrong and criticize them for it. To our employees and to ourselves, our roles as bosses hadn't fundamentally changed. It was now clear that they would have to.

PAINFUL TRUTH

In June, our management team faced yet another crisis. Cable orders had dropped so low that we decided to short-time the shop, shutting down the operation for two weeks and sending all employees home on a temporary layoff. The planning organization was still talking to me about closing the whole copper operation and transferring all production to Phoenix and had actually begun work on a formal proposal. It was at this point that I made what would turn out to be a colossal mistake: I kept this information a secret. While Walt and Lee knew, I'd been keeping the situation to myself in my own organization, not even telling my Copper Shop managers, and now that the onset of doom was even closer, I decided to leave it that way. I felt that I wanted to protect them from such terrible news; they already had enough to worry about, and, besides, what more could any of us do?

Closing the copper operation would affect the lives of almost 1,000 people. Many would lose their jobs. This thought began to occupy my mind almost constantly and put me into a serious personal crisis. I felt responsible and became gripped by a nagging emotional pain that wouldn't let go. I felt that my autocratic management style was at fault. I and all the other managers would have to change if we were going to survive. We needed everyone to pull together, a real team approach, but I didn't know how to make it happen. I agonized over this issue constantly, trying to see a way out of it. I couldn't believe the situation was real, that it was actually happening. What had become of the good old, stable, worry-free Copper Shop? And despite my depression and frustration, I couldn't bear to tell anyone this horrible news; the very thought of bringing this information to my team made me sick with anxiety. Figure 2.6 in Chapter 2 illustrates the cycle I had slid into.

This was a very lonely time. I'd been coping on my own with the idea that the shop might close; no one on my team knew about it. How, then, were they supposed to support me as a team? I was sending lots of messages about teamwork, quality, and sharing success stories, but the positive feedback was almost nonexistent. Although I strived to shift our focus to quality, quantity and efficiency continued to be the operation's primary objectives. I think fear paralyzed my managers and kept them from changing, from doing their very best. And I was burning myself out from the lonely isolation of keeping the Copper Shop's precarious fate to myself. Well, it was time for me to "put up or shut up." The wonderful energy from the Deming seminar seemed faint and far away, but the learning was still there, and I wanted to act.

CHANGE

The moment of truth came in late June at one of our weekly staff meetings. There was a lot of bickering, complaining, and finger-pointing, all of which I'd been hearing for weeks, and not a single comment was unfounded. Nobody was complaining without reason. I listened to all their frustrations, all their concerns, all the reasons why things were so fouled up, and realized that this was exactly the right time to unite the management team.

I slammed my hand down on the table, creating a considerable "BANG" in the room. Conversations stopped cold. There was dead silence, and I had everyone's complete attention. I began speaking slowly and softly, with firmness and sincerity, and told them my secret: the Copper Shop was within a hair's breadth of being shut down. But, there was a chance we could save it. Within seconds, everyone had become focused like I hadn't seen in a long time.

We agreed that we couldn't save the shop alone. Everyone jumped on the idea of building a unified team encompassing all our people: supervisors, engineers, and production specialists. Involving the production specialists was vital. There could be no meeting of the minds without their cooperation and support, so we decided to approach them with an honest plea to help save the Copper Shop; we hoped to unleash their creative energy. As Deming had made so clear, these people were the key. They were the Copper Shop. They worked there, knew the problems, and were in the best position to provide solutions.

With a bang of my fist, I moved into the "try again" cycle shown in Figure 2.6 in Chapter 2. Dr. Deming had been my Coach, but my staff had become my Mentor. The key was to share my problem with someone who cared, and accepted and supported me. This is not "career mentoring" by senior management but "personal mentoring" by subordinate colleagues.

At that meeting, I offered a direction built around a few simple principles. First, we would work as a team, no longer focusing on individual department efficiency but instead measuring our effectiveness as a total organization. Second, we would measure our success by the quality of our products and processes, not by financial performance. Third, we would each be responsible for our coworkers' success, and they in turn would be responsible for ours. In summarizing, I said, "I won't ask very much of you. I just want you to do the impossible. And every time we have a meeting, we will share success stories, no matter how small or insignificant they may be. Our focus will be our people and the impossible things they accomplish." My job began to get very interesting.

ACTIONS SPEAK LOUDER

We started by going out on the shop floor and spending an enormous amount of time with our people, but with different intentions and feelings than before. We stopped blaming people for problems. Instead, we looked for ways to support them, to take care of their needs so they could take care of our customers. Instead of looking for what they were doing wrong we began looking for what we (management) were doing wrong and providing the necessary support for our people to perform their jobs successfully.

One of the toughest hurdles we had to cross was creating an atmosphere in which people felt comfortable stepping forward and telling us what was wrong. This became my personal focus: building the team and driving out fear. We convinced people that once problems were identified, they would be fixed and not covered up.

We told everyone it was okay to make mistakes—that mistakes are a natural part of life and we all learn by trying.

It wasn't automatic or easy for people to open up, to express themselves and to disclose how things weren't working. For years, we had discouraged real honesty; people who had been too open and who had the habit of taking risks and exposing problems were ridiculed, punished, or shunned. Those who were the first to open up in this new quality initiative were real risk-takers.

During the next three months, my resolve would be tested dozens of times. At first, I struggled with each decision, which usually dealt with major quality issues. There were still many questions and pressures from top management about our financial performance. There were times when my patience grew thin, but over time our managers accepted our new management approach. They started making the tough calls they previously had asked me to make. They became believers.

Quality became the glue that held us together. By October, we had formed quality teams on all three shifts, consisting of production specialists, engineers, and supervisors. We also formed task teams and cross-functional teams to work on special problems. Gradually, our teams created a widespread change in attitudes and helped us work together as partners instead of adversaries. At meetings, we spent more time sharing success stories. This gave us many opportunities to express appreciation for people who were supporting one another and making things happen.

Building on the team concept, we began structuring all our internal interactions to model customer-supplier relationships. This meant that the first operation in the shop was the customer for an outside vendor, and, at the same time, was the supplier for the second operation in the shop, and so on. Each internal customer had the right to set the specs for each internal supplier's product, and suppliers were responsible for meeting those specs.

At first, it was difficult to get the engineers and supervisors to accept ideas and feedback from the production specialists. Our managers and I attended many of these early meetings to show support and sometimes to encourage the supervisors to invite the production specialists into meetings. We discussed ideas that were tried but didn't work, and at such times always recognized the value of the people who had tried. Many times, when we reviewed a failure new ideas would emerge. We would build on these, consistently trying again until success was achieved. Little by little, the mood of the Copper Shop began to change, as people at all levels of the organization finally began to get the message: we wanted them to be open, honest, and resourceful—no joke.

In early October, I attended a team meeting at which a production specialist named Mark McCullough presented some ideas. Although he was a little nervous and had no experience in such a situation, he produced three pages of things to fix; I was delighted—this was just what I had wanted to see. But each time he mentioned something, despite some appreciative discussion, it never seemed as if anyone planned to take any action—no one even recorded his ideas. I felt we might lose an excellent opportunity to show this team we were serious, so after the meeting I asked if I could make a copy of his hit list. He smiled and said, "No, it's not complete. I'll work on it at home and give it to you tomorrow." When the team met the next day, Mark produced five hand-written pages of action items! This time, by the end of the meeting we'd assigned every item to members of the team. We were off and running, starting to make it happen. I left the meeting with a stronger sense of optimism than I had felt in months.

Mark had simple, creative, and cost-effective ideas for solving problems, some of which were the very soul of design elegance. His "move the hole" solution is a case in point. Mark's department used machines that twisted thin insulated wires into pairs and then assembled the pairs into larger finished cables. These cables could be up to 50,000 feet long and involve up to 3,600 pairs. If even one wire in the finished cable were out of spec, the whole cable would have to be repaired, which can be a very expensive operation.

There were 25 different color-coded twisted pairs of wire, and each pair was identified with a paper tag. Each tag was attached by slipping a wire through a hole punched through the tag about half an inch from the edge. The problem was that the tags tended to tear through this thin margin of paper, and the identity of the corresponding wire was lost. Mark's solution: put the hole in the center of the tag instead of near the edge—hmmm. We implemented the idea immediately, and the improvement was quite dramatic.

Altogether, Mark's ideas resulted in more than 25 improvements to the operation.

During this period, the great irony in our situation dawned on me. Mark's resourcefulness and ingenuity, and that of his coworkers, was nothing new. They'd always known what was wrong, but when they tried to tell us, we never listened, and most of the time they were afraid to tell us anything anyway. What a waste, what a tragic underutilization of talent—and it was our fault.

By the end of November, the word had spread that we were really serious, that this wasn't some flash-in-the-pan quality initiative—it was real. At regular meetings, we would spend the last 15 minutes sharing team success stories. At first, it took the managers' involvement to make this work. A manager would either tell a story or encourage someone from his department to speak up.

Many times, a member of our team would talk about a project that hadn't produced any result. He would share the problems, the struggle, and the associated pain. Encouraging this sort of story was our way of letting people know that trying something and failing was still a success. And at the end of each story, we always applauded.

In time, our production specialists, supervisors, and engineers were enthusiastically standing up and sharing their own success stories, with no coaxing from their managers. Often, a story became a way for a person or team to say "thank you" to other members of the Copper Shop for helping with a problem or just providing moral support. It became a way to say "I like you and thanks for caring." As the momentum increased, our production specialists were asking for more responsibility, and we gave it to them.

Toward the end of 1985, we started a series of quality forums for all employees. The forums were designed to increase their awareness of business issues and provide an opportunity to speak openly about their concerns.

The forums brought the various levels of the organization together in a nonthreatening setting. A typical meeting would include 25 people representing a full cross-section of the population. Their meeting would last about six hours and was facilitated by a member of the training organization. A member of the Atlanta Works management staff joined the group during the last two hours to discuss issues that had come up in the first four hours. We wanted to treat all employees equally, so the format was consistent on all three shifts. The forums served an important purpose: open communications. People tend to panic during a crisis, rumors fly, morale drops, and everyone worries about job security. The forums gave us a clear channel of communication and helped all of us to see a way out of our problems.

ROLE MODELS

We were learning a lot. We were growing and improving in remarkable ways, but there remained many tough business problems to solve. There were still fires to fight and external customers to satisfy, and the omnipresent planning organization was calmly inspecting their financial reports to convince themselves of our viability. While the Copper Shop had made tremendous strides internally, the cold, uncompromising numbers describing our economic performance had yet to improve.

At my urging, many of the other managers had attended one of Dr. Deming's seminars. One of the new faithful was Hub Evens, who returned from his Deming seminar in September just as inspired as I had been, if not more so. Hub shot back to work like a tornado, eager to apply the Deming method in his department. He met with his people to share his thoughts on the changes taking place on the job and his ideas for a better way to run the business. He told them how he planned to drive out fear and build a relationship founded on trust. This is all great stuff, right out of the seminar, which Hub and I had been talking about for months and which other managers were already applying with great success. But this time things were a little different.

After about 30 minutes of explaining all his great plans, Hub called a break and left the meeting. He didn't feel right about what he was saying, because his recent actions didn't support his words; a considerable credibility gap had opened up and swallowed him whole, and he knew it.

While things were beginning to improve at the Copper Shop, we were still in a considerable state of crisis, and fire-fighting was still a part of our regular routine. In the face of a crisis, old management habits tend to be the rule of the day, even for people who are trying to do things differently—and here was the problem that had stopped Hub in his tracks. A few days earlier, he had suspended two workers for three days without pay for covering up a manufacturing defect. That was before the seminar. Now, he realized the workers probably did what they thought management really wanted: if it's good enough, ship it. Although we were saying "quality first" all over the place, they simply didn't believe us. We had always preached quality in the past, and if it was so much hot air then, why should they believe us now?

Hub had acted according to conventional wisdom. With his punitive action, he was trying to send the message that it was vital to produce quality. But the message the workers got was the same one they'd understood for years: "If you make a mistake and get caught, you'll be punished, so you'd better not get caught." Hub had a major credibility problem on his hands, one of the toughest things a manager can face.

Hub phoned Walt and me immediately and asked if he could bring these men back and pay them for the days they'd been out. This was a novel idea, to say the least, and we'd never done anything like it before. But we knew it was important to support his recommendation. We had realized it would be important for us to talk openly about our mistakes and to encourage others to do the same. For virtually everyone in the Copper Shop, the kind of openness on the job we were after was a strange and puzzling behavior. It wasn't always natural for people, so we had to become role models. For Hub, it was time to practice what he preached. He returned to the meeting and admitted to the entire group that he'd made a mistake.

The same day, Hub and I called the two men back and told them they'd be paid for the time off. We met with them separately, apologized for the punishment, and explained why it was so important not to cover up any errors, including our own.

During the meeting, we explored why they had shipped marginal product and discovered that they believed we were more interested in shop efficiency than quality. They never thought what they did was wrong or that we would be upset about it.

The first man shared his story, saying, "You've given me back my dignity and my pride. I didn't have any more vacation time and my kids knew this. When they asked why I was home it was very painful to tell the truth. I told them I made a mistake at work and didn't do the right thing and was being punished. I did something that I had been telling them not to do." I was deeply moved by this man's candor, by his simple, frank, and emotional honesty.

The other man had a very different story. He said, "I was going to get you. I was going to get even, but I can't do that now." He was serious. I believe he would have gotten even somehow, if we hadn't admitted our mistake. It made me wonder how many times workers had exacted some sort of revenge in the past. We all left that meeting feeling inspired, energized, and hopeful. Inadvertently, these production specialists had done a lot to help us become better managers; and by the way, both of them became much better performers in their own work.

We didn't realize how significant this decision would be. Later, we learned that this event was profoundly symbolic for the entire Copper Shop. The simple act of calling those two men back, apologizing, and paying them for time off represented something concrete, real change, new opportunities, and a brighter future. It carried enormous power. The story spread through the plant within hours, and this time people got the message we intended: "We care about our product, we care about each other, and we're all on the same team."

When our management peers in other organizations heard we had paid some employees for time not worked, they thought it was a foolish and dangerous decision, especially since we were having poor financial results that were being analyzed by the planners. They were deeply concerned about the precedent it would establish. They never acknowledged the magic that took place on the shop floor, where the people saw it as the first really significant caring act and that the Copper Shop management was serious about doing things differently.

INTO THE FIRE

Just about the time things started looking up, we were hit with another bomb: upper management had decided to expand our fiber optic operation, again. This meant, for the second time, the copper operation would have to pull out machines to make room for new fiber optic equipment.

We agreed to remove a bank of older wire-insulating machines that didn't produce the daily production rate we needed anyway, and as such could vanish with no real impact to our financial picture. Unfortunately, these machines were not in the location best suited for the fiber optic expansion. I didn't think this mattered too much, but apparently others outside the Copper Shop thought it mattered quite a bit. A decision was made to pull out not our oldest and slowest insulating machines but our newest and fastest. It seemed that our best machines were occupying the best space for the fiber expansion, so they would have to go. This action created a new and even greater sense of crisis, not just in the insulating department where the immediate impact would be felt, but throughout the entire Copper Shop.

We fought the decision as best we could, but we lost. Pulling our best machines seemed foolish, counterproductive, and crazy, especially in light of how hard we'd all been working. It sent a chill of fear throughout the shop and crystallized our deepening sense of panic. For many, our crisis finally became tangible, concrete, and no longer just a rumor. Was this the beginning of the end of the Copper Shop? We all saw how easy it would be to lose our jobs.

OUT OF THE ASHES

But a wonderful thing happened. Instead of being the final, deadly blow to a wounded organization, this crisis turned out to be the final push needed for us to leap over all remaining obstacles to quality. Suddenly, you could feel a very high energy level pouring from everyone in the Copper Shop. Supervisors on the shop floor who previously claimed that quality came first now visibly supported their claims with actions. Operators were empowered to shut down machines and have them fixed if there was any question about the quality of the product being produced. We became highly focused on satisfying the customers. The concept of internal customers and suppliers was taking hold and working better than we hoped it would.

We all kept trying different ideas: some worked, others failed miserably. Constant course corrections became the norm. Without any formal agreement or process, people created what might be called a communal brain trust, sharing information about what worked and what didn't. If one department tested an idea and it worked, they would share it with another department where different people would try it. Each time we had a success, we would celebrate. We established a routine of recognizing people who had tried something and failed and gave credit to those who were trying their best.

The managers had their own ways of encouraging people to bring problems fearlessly to the surface. Hub instituted his popular "Yellow Flyers," each prominently identified with the headline: "Problems Exposed." These were simple, letter-size bulletins on yellow paper that gave people recognition for finding mistakes. An example of a Yellow Flyer is as follows:

Thanks Carl. Carl, a production specialist in rewind, brought to the attention of his supervisor the fact that his machine had a cracked sheave. Carl called maintenance and had the sheave replaced. He questioned the quality of three reels previously run. He checked them, found two defective, and scrapped them. All of our customers thank you for your attention to detail. It's people like you that will make this company go.

The flyers were also used to express gratitude to individuals who had done a particularly outstanding job:

Congratulations Rewind Teams! In the past week the quality of your repaired wire was fantastic. The cable made with your wire tested almost perfect, only one defect found. Your attention to detail produced the results described. We all thank you!

And sometimes, people were commended simply for pointing out a problem, even when they played no part at all in its solution:

Congratulations Quality Improvement Team! During a recent insulate quality team meeting Mary and Billy pointed out two quality problems; the first had to do with the cooling water and the second with the annealer quench water sumps. Both could cause capacitance and wire-break problems. The team determined cleaning schedules had not been followed. Maintenance was called, the systems have been cleaned and a check process has been implemented. Thanks to your attention to details, constant improvement to quality will be achieved.

Hub distributed these flyers to people in his department, posted them in the shop, and shared them with other members of the management team. The people acknowledged in these flyers did not necessarily solve the problem they discovered, but they had seen a problem and let somebody know about it. This idea of openly and publicly admitting the existence of problems was a major change in our culture. Changes like this created an environment where people felt safe to express their honest thoughts and behave in accord with their deepest feelings. The fear that had eaten at our strength for so long was melting away at last.

Hub's Yellow Flyers, and other policies like them, created excitement and team spirit on the shop floor. Everyone seemed to have one common objective: to improve the results at our final test operation, the point at which the product is checked before it's shipped to the customer. Everyone in the Copper Shop got into the act, including maintenance, production control, accounting, personnel, training, and purchasing. The team idea spread faster than we dreamed possible. Most of the time, there were so many projects in the mill it was impossible to know who was doing what. In many respects, as managers, we were actually no longer in control of the operation, and things were running more smoothly than ever. We loved every minute of it. And I finally learned the most valuable lesson from this whole series of events: good management requires letting go.

It was exciting to come to work. Each day brought a new adventure. Instead of an endless journey, it felt more like a series of short exciting trips, each one ending with a celebration.

INTO THE FUTURE

Almost 18 months had passed since our first crisis, and the improvements were starting to show everywhere we looked. Our finished product test and inspection results showed a quality improvement of 13% and were still getting better. The need for rework and retest had dropped off dramatically, resulting in a greater than 40% work-in-process inventory reduction. A manufacturing cycle time reduction of 20% let us improve our on-time delivery of customer orders. These were significant, measurable accomplishments that took all 1,000 employees pulling together with a common goal. We no longer looked at shop efficiency as an important measure but were now focused solely on delivering high-quality products—the best that had ever left the factory. The ratings for our product quality had returned to a high level, and our manufacturing costs were coming down.

The people in the insulating shop, who had lost their best machines during the last fiber optic expansion, reached a production level of which my managers and I had only dreamed. These people were amazing. When confronted with the crisis of losing their best equipment, their reaction was clear-headed, honest, and sincere. They understood the seriousness of the situation and responded by asking what needed to be done to guarantee we wouldn't go out of business. They even set their goal higher than what we had asked for. Using mostly older and slower machines, they actually set a new record for machine output—an increase of 18%. When they broke the old production record, we celebrated and thanked them for their support and commitment.

I learned a great deal from the people in that department. I learned that when people feel a unity of purpose, participate actively in setting their objectives, and drive themselves by their own determination to be the best, they can do what at times seems genuinely impossible. That's a funny thing, what's possible. I realized that the assessment of what is possible based on traditional systems of management is often not very realistic at all.

During the previous year, we had done the impossible because we had learned that mistakes were stepping stones to achievement. Traditionally, like many people in business, we had always looked for the single right answer to fix a problem, trying to discover the perfect solution fully formed with all the necessary details.

In the past, if an idea didn't seem perfect, we'd reject it. We had never approached problems with an attitude of creativity or experimentation. Trying out a solution that we knew had some flaws mixed in with the virtues was simply not an option. But as time went on, we began to recognize that the best problem solving involved building from many small ideas. Each idea would inch us along, slowly adding to our success. We'd take a piece of this one, a little of that one, add them both to a new one from an unexpected quarter, and gradually the ideal solution would emerge. This is quality.

To many people, quality is limited to statistical charts, high-tech machinery, and automated processes. But a group of managers, engineers, supervisors, and production specialists in the Atlanta Works found the meaning of quality in human terms: teamwork, appreciation, caring, and sharing. More than anything else, the power that transformed us was a shared sense of crisis. It wasn't until we held a focus group session with Marilyn in August 1986 that we realized how strong our bond had become. For many, we were more than a team—we had become family.

Our Copper Shop team played their hearts out. Many days it felt as though we were in some sort of tournament finals, fighting to hang on. In the beginning of our 18-month journey, our mood was bleak, but by the end we had done everything that was needed to survive and we actually prevailed. Over time, our internal links as an organization had radically changed. Our emotional bond was our gold medal.

If there's a secret to our success, it must have had something to do with our people. Igniting the emotional spark by caring about each other's feelings was the real turning point. A manager must believe in the people, must know the people, not necessarily the details of their lives, but their hearts and dreams. We must be connected. We must care.

THE LINK BEGINS

While I had been sweating, learning, trying new things, sweating some more, and finally celebrating our success in Atlanta, Marilyn had been hard at work in New Jersey developing the Quality Leadership Conference. The QLC was to be an executive education conference on business quality and was intended to spearhead the Network Systems quality initiative for the coming year. As mentioned in the introduction, the Copper Shop story had been selected from more than 200 stories, after review by the QLC planning team, because it was in sharp contrast to all the others in its focus on people. Virtually all the stories involved people in one way or another, but the applications submitted to the team for review didn't focus on people. The stories, as told in these applications, seemed to add the human element as an afterthought, as if it didn't really matter.

Marilyn wanted to meet with some of the people who had lived the story to discuss plans to document it. Ultimately, she wanted to produce a video module for the conference—one that told the story of our success and identified the elements responsible to inspire similar breakthroughs at other places in AT&T. I was delighted, but in the spirit of trust and empowerment that was now the life blood of the Copper Shop, I had to tell her that if the people involved in the story didn't want to participate, I wasn't going to force them—it would have to be a group consensus. If Marilyn wasn't convinced before, this must have brought it home to her that we were really doing things differently.

Marilyn came to Atlanta on July 25, 1986, and presented her ideas to me and 15 team members; together we represented all levels of the Copper Shop. She described the QLC in great detail and discussed why our story had been selected, going beyond the general explanation that it was the only one with a human element. Our story demonstrated other factors that Marilyn felt were quite important: insight into a problem, persistence in overcoming barriers to its solution, and team spirit that led to increased mutual respect throughout the organization.

The atmosphere during our meeting was energetic, warm, and open. We told stories and tossed ideas around, and finally it came down to a vote: did we want to participate in this thing or not? We all agreed to continue, and Marilyn made the arrangements.

EXPLORATION

The next step was a focus group meeting on August 8th at a hotel in Atlanta, which brought the Copper Shop people together with Marilyn and Bill Idol, an international management consultant and expert facilitator. Also present were a script writer who would work on the video and an A/V engineer to record the proceedings.

Bill Idol explained the two goals of our session: to provide the script writer with enough information to produce a successful video and to develop a better understanding of how our success had been achieved. Bill was terrific: friendly, warm, and clear. Within minutes, he had won our trust completely. We had no trouble speaking candidly about our experiences with the work and each other.

The second goal was Bill's real focus. He wanted to consciously recreate our success, uncovering every element of the process. Clarifying the reasons for our success would mean that even if the video were never produced, all of us would still learn exactly how our success had happened. We'd understand it well enough to replicate it in the future and would never have to stumble across it again.

Bill structured the session to be open and unintimidating. The room was fairly large and the atmosphere informal. We sat two at a table in a large semicircle, but he also encouraged us to wander around, get coffee, and generally be as relaxed as we wished.

He started by having each of us write down the events that we felt were important during the past two years. This was just to give us some focus, and he only allowed a few minutes for it. Next, we began creating a rough timeline to explore the sequence of events in the Copper Shop. After all the discrete events were in place, we began to fill out the cast of characters, the people who featured prominently in the story, and the sort of roles they played at different times.

During this process of recreating our story, I was impressed by Bill's willingness to accept different perspectives of what had happened in the past. He made it clear that this was not some sort of inquest and that it was not our job to reach a consensus. He was interested in our individual perspectives.

He emphasized the idea that any one of us might have some unique perception that could turn out to be really valuable, even though it may seem out of step with the mainstream. He wanted no opinions or feelings hidden.

Bill also wanted to know about our perceptions of change during this period. What did we see happening? How did we know a change was taking place and whether it was real or imagined? This helped identify key milestones in the timeline. We also suggested scenes for the video that would help viewers understand the story.

For the next several hours, Bill collected everyone's recollections of events, people, and perceptions that led up to where we were. He allowed time for us to compare perspectives and explore events more deeply. He would often ask very probing questions and summarize what people had said to be sure he understood. He encouraged us to talk openly about our fears and feelings, keeping us on track the entire time.

Most of the time was spent talking about our pain, our failure, our struggle to survive, and the many mistakes we had made; almost everyone focused on the negatives. We learned that our sense of crisis and the many impossible challenges we'd faced had forged a strong emotional bond. As people poured out their stories, showing more and more feeling as time passed, we eventually reached a level of emotional communication that we'd never quite managed on the job. We had come close, but we'd never taken the time to tell each other what the past months really meant to us. We were always too busy or too uncomfortable to be openly emotional on the job, to reach out and either give support or take it.

Telling stories and recalling distant moments in this safe, protected setting made us open up, and we needed to. We shared events that were relatively public as well as some that were very personal, and a few stories were a surprise to almost everyone. This experience was both a rebirth and an awakening. We felt the memories of things past but also began to see a new structure unfold as Bill recorded more and more details as the team revealed them.

The focus group ended with some written feedback and evaluation at about 3 P.M. Everyone who attended was on a natural high—recreating two years worth of emotional turbulence was a very heady experience, and the attention our story had received during the focus group was like the cheers of a crowd. Bill promised to work the entire weekend to complete his report.

SECRETS OF OUR SUCCESS

We received it early Monday morning. While the results were basically what I expected, Bill had shaped them into a set of insightful and elegant conclusions with far-reaching implications. Like a scientist searching for a new vaccine, he had isolated what he felt were the two key factors that accounted for our success.

The first was a genuinely shared sense of crisis. Bill noted that during the last two years, we had lived through many individual crises, beginning with divestiture itself and getting worse as time passed. This perception was obvious to all of us, but Bill had seen a relationship most of us had missed: it was precisely this sense of crisis that had provided the fuel for our recovery. While the crises themselves were horribly negative, without them there would have been no positive turnaround and no miraculous success. Our story showed a clear causal relationship between crisis and opportunity, which can be used to advantage only when management recognizes its value and takes appropriate action.

The second factor was a conscious change in management style from controlling to empowering. Most of us had essentially stopped being managers in the classic sense and had started being caring leaders. This change grew from the top down and had to be consciously reinforced over many months. Only very gradually did the belief in empowerment, and the corresponding willingness to let go of total control, spread through the shop's entire management team. Bill saw a familiar friend at the heart of this change: Dr. Deming. Deming knew that driving fear out of our organization, an effort that must be consistently lived and reinforced by management, was the key to successful empowerment.

Bill identified four phases in our transformation, resulting in a remarkably simple pattern:

1. management living quantity—quality only words;
2. management begins to live quality;
3. employees accept and begin to live quality; and
4. quality is alive and growing.

This pattern was immediately understandable and recognizable as the process we had experienced. Considering these four steps, with no effort at all I could think back to individual moments that characterized each one perfectly.

I didn't know it yet, but there was also another structure at work, another pattern behind the Copper Shop story. Some time later, Marilyn called me in a terrific state of excitement and began telling me about something called the American quality archetype.

My awareness of the past two years' events had already been enhanced by Bill Idol's analysis, but with the new information that Marilyn was pouring out, the Copper Shop story began to take on the quality of a mythic journey or an icon.

A TEMPLATE FOR QUALITY

I know now that, whether we see it or not, the American archetype of quality is everywhere. It exists in government, business, and education. It drives the interactions of our military, our children, and our presidents. And it was the basic motivational force at work in the Copper Shop between 1984 and 1986. There follows a comparison between the archetype structure, as described in the previous chapter, and the events of our story of failure and success.

Phase I: Crisis and Failure

At the Copper Shop, we experienced three Lawgivers: AT&T's competitors, the new fiber optic technology, and the management decision to phase out our operation if our performance didn't improve. Each of these factors precipitated its own set of crises.

We felt the first crisis through the loss of business to our competitors. We had just gone through divestiture, and the Regional Bell Operating Companies, now free of their corporate link to AT&T,

often found it convenient to purchase cable from other vendors. This loss of business had an immediate impact on our operation.

There was also substantial pressure to bring the new fiber optic technology on line. Fiber was the future, and it was vital to become competitive in this field as quickly as possible. When they needed room to expand fiber optic production, they simply reduced our copper production capacity. Several times during that 18-month period, some of our copper cable machines were removed, and our best insulating machines fell to the planner's pencil in late 1985. We were horrified—a destitute family watching the bank's repo men haul away their furniture could not have reacted with more grief.

This succession of experiences made the crisis tangible, and the last one gave the final necessary push. It had the emotional impact that forced us into action, creating the energy needed for everyone to change and pull together. Once I revealed the pending management plan to close down our operation, it added more fuel for the crisis. During the focus group, Bill Moore, shop supervisor, expressed it better than anybody. He said, "I felt we were treading water with one nostril above the surface." When I heard that, I remember visualizing myself completely submerged, definitely drowning. It was a difficult time for everyone.

For a very long time during this period, I was closed to learning. As manager, I mirrored our organizational crisis with a personal one. I eventually realized that if we were going to save the Copper Shop and the jobs of 1,000 people, I would have to change personally. I'd made a mistake keeping that management plan secret—it would have been far better to share it with everyone. It would have created the shared sense of crisis sooner and probably would have resulted in our taking action much earlier. At the time, it seemed more important to protect people from pain. I didn't realize that if we were going to get our energy flowing, they needed to feel the pain as I felt it. Figure 3.1 illustrates Phase I.

Phase II: Support

As time passed, I kept looking for a better way to run the shop but couldn't find it. It wasn't until the Deming seminar that I found a new purpose, a determination to build a team around open and

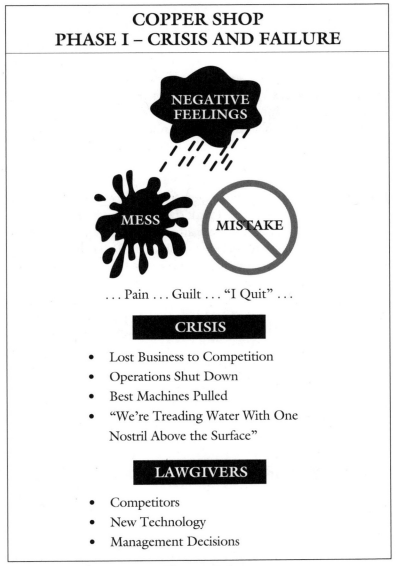

**COPPER SHOP
PHASE I – CRISIS AND FAILURE**

NEGATIVE FEELINGS

MESS

MISTAKE

. . . Pain . . . Guilt . . . "I Quit" . . .

CRISIS

- Lost Business to Competition
- Operations Shut Down
- Best Machines Pulled
- "We're Treading Water With One Nostril Above the Surface"

LAWGIVERS

- Competitors
- New Technology
- Management Decisions

Figure 3.1

honest relationships. Until then, I would blame my management team and the production workers when things went wrong. Through his teaching, Dr. Deming became my Coach.

Two of Deming's points really made an impact: management is responsible for 85% of the problem, and workers live in fear. If you want people to do their best, remove the fear. After my managers and I began to apply this learning, we started to approach managing the business with a style very different from the threatening autocracy of the past. We tore down barriers and built bridges to a team of people committed to quality. We didn't realize it at the time, but a fundamental change in our mutual respect took place.

Little by little, we became open to learning. We learned to support one another, to listen, to understand. Managers became Mentors, allowing our people and me to try new things and fail without embarrassment. My managers were my Mentors and our peoples' Mentors. When we failed, they would reaffirm our worth and encourage all of us to try again. Hub's Yellow Flyers were an excellent way to give risk-takers tangible, public recognition. The shop supervisors and a number of production specialists provided the day-to-day coaching we needed to make the team structure work. Instead of getting work done through people, as in the past, we were now nurturing and enriching our people through work.

Before this transformation, we were always fire-fighting. In that mode, you hide your problems, or "red beads" as Dr. Deming calls them. You cover them up and put bandages on them. Now we were going after the red beads and getting rid of them instead of just hiding them. We were putting real solutions in the system, taking the time necessary to fix problems, correctly and thoroughly. Figure 3.2 illustrates Phase II.

Phase III: Celebration

During the 18-month journey, little celebrations were going on all the time, acknowledging the latest Champions throughout the Copper Shop. We gave frequent recognition to individuals and teams of people who were doing the impossible. My favorite is the story of the supervisor who shut down his repair operation for four days, until he and his production specialists were convinced that the engineers and maintenance people had fixed their machines properly—they would accept nothing less than 100% reliability. He and his people refused to send any questionable product to the next team

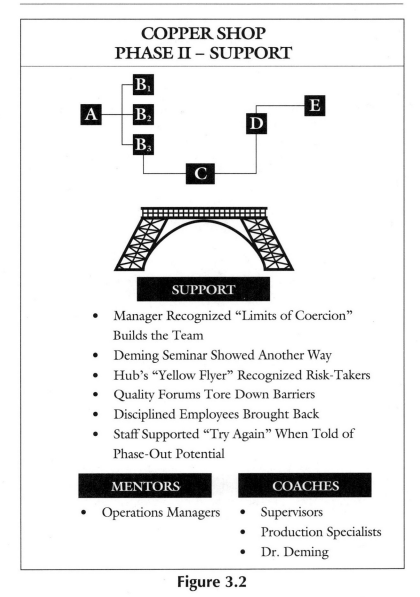

COPPER SHOP
PHASE II – SUPPORT

SUPPORT

- Manager Recognized "Limits of Coercion" Builds the Team
- Deming Seminar Showed Another Way
- Hub's "Yellow Flyer" Recognized Risk-Takers
- Quality Forums Tore Down Barriers
- Disciplined Employees Brought Back
- Staff Supported "Try Again" When Told of Phase-Out Potential

MENTORS

- Operations Managers

COACHES

- Supervisors
- Production Specialists
- Dr. Deming

Figure 3.2

in the cable-making process. He was scared when he did it, but when we publicly applauded his effort and showcased him as an example of what needed to be done, he felt proud.

Each department found their own way to celebrate, but we never took time out to celebrate the entire Copper Shop accomplishment. In fact, we were so busy keeping the operation afloat that we never realized we had stopped sinking. It wasn't until after the focus group session that we saw how important the magic ingredients of quality and caring were to our success.

The closest we came to having an emotional celebration for the whole shop was the management appreciation breakfast held by the Plastic Insulator Shop production specialists in March 1986. They paid us a tribute for a job well done and made complimentary remarks on behalf of the group.

When I stood before the group, ready to say a sincere and heartfelt thank you, I had a lot of trouble finding the right words and felt a sizable lump growing in my throat. When I started to speak, I was surprised to feel a tear rolling down my cheek; well, the last two years had seen a lot of firsts in my life, and this was the first time I had ever cried at work. No one laughed. For the first time in my career, I had felt the heartbeat of our people, and it was the best workday of my life. Looking back I now realize this was the time when our worth as a team was reaffirmed.

When the focus group finally met, the secure, private environment let us share things that none of us would have confessed at work. During those moments of telling truths and sharing secret feelings, we related with each other more intimately than at any other time. When the session ended and we came together to say goodbye, something special happened: instead of shaking hands, we embraced. We had become much more than friends—we were a family, people who would always share a precious moment in time. We didn't realize it then, but the focus group session had been a powerful celebration. Phase III of our story is captured in Figure 3.3.

Jerry Butler, a production specialist in the Copper Shop, summarized our evolution beautifully, with wonderful insight and understanding:

I think of us as being a family, not just a team. The way I look at this thing, we were single. And now we've entered into a special relationship, almost like marriage. It's a commitment, not just to our customers but to one another. When we

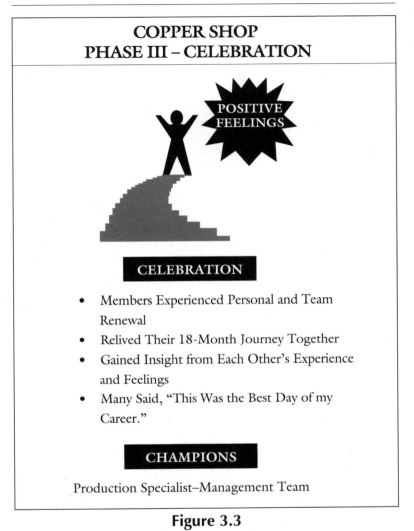

Figure 3.3

began to care about each other's feelings, that's when the turn-around started. What we're talking about is making a product work . . . satisfying our customers, so that they come back for more and more and more. That's the way I sum it up. Now, the marriage isn't perfect, but it's growing, it's getting there, caring about one another. Care makes it work.

On my feedback form after the session, I echoed Jerry's thought with a comment of my own and wrote, "I feel better than I ever felt before … I thought we had built a team, and now I know we have a family."

"CARE MAKES IT WORK"

In January 1988, I received a remarkable piece of news from a very excited Lee McClary: the Network Systems president had invited Lee and his team members to the corporate headquarters in New Jersey to accept an award for their outstanding quality improvement. As part of the award ceremony, there would be a premier showing of the video "Quality … Care Makes It Work," documenting our Copper Shop story. Interviews for this program were taped at the factory in September 1987 and featured the comments of production specialists Jerry Butler, Mark McCullough and Larry McDaniel, supervisors Bill Moore and Doug Rustay, operating department manager Hub Evens, engineering department manager Don Reser, director of manufacturing Walt Ehmer, manufacturing vice president Lee McClary, and me.

When I spoke with Lee about the award ceremony, it was clear he'd already made up his mind not to attend. "Lew, it's your story and I want you to decide who attends," he said. "Just remember, the team selected will represent all the employees at the factory."

My decision was an easy one. I would select the people who were the stars of the Copper Shop video; after all, it was their story, and no one could tell it better. I knew their passion and enthusiasm would win the hearts of the top management at corporate headquarters.

A few weeks later, we were all at the airport, ready to catch the red-eye flight to Newark. We were all tense about the pending activities at corporate headquarters and were doing our best to hide it. Before long, Larry McDaniel quietly admitted a more immediate concern: this was his first flight. That's all it took to break the tension. Hub, Walt, Bill, Mark, Jerry, Doug, and I spent the next 45 minutes reassuring Larry. It was a good distraction for the rest of us and took our minds off our own nervousness. After boarding, we told the flight crew that this was Larry's maiden voyage. Much to our surprise, the pilot came back to talk to him. He told Larry not to worry about a thing and that he and his crew would make this a special flight. We all laughed and felt more relaxed, and I took some pictures of the pilot and Larry for our scrap book. A few minutes later, we were airborne.

From the moment we arrived in New Jersey, we were given the red carpet treatment. A special greeting at the airport, a limo to take us to headquarters, flowers for our lapels, the video premier before an audience of executives, and a special luncheon attended by the Network Systems president.

This experience was something special for all of us. We were all shocked at how open the executives were with us and how easy it was to be open with them. All the members of our "A Team," a name with which we became identified, believed we were on a mission. This was our opportunity to be spokespersons for all our coworkers in the Copper Shop and to share our story with decision makers in the corporation with the hope it would leave a strong imprint of the changes that people can achieve.

We all carried the message, but production specialists Larry, Mark, and Jerry made the biggest impact. When asked to explain what was responsible for the change, Jerry said, "The caring was the key. Without management allowing us to be creative, we couldn't have done it. That's what the video is all about. We're not only expressing what we went through but what our coworkers went through." Jerry, Mark, and Larry agreed that the turning point was when management asked for help and began consistently backing up their words with action. Larry said, "Quality had to be number one." If it wasn't first, we'd be in trouble." Supervisor Bill Moore added, "Operators developed a strong commitment to quality production." The team agreed that the message took months to take hold, but over time, management's candor and willingness to let people flag problems without fear of retribution made a lasting impression.

For the eight of us, nothing would happen that day that wouldn't be remembered for a lifetime. Each of us would recall a different event as the highlight of the day's celebration. For me, it was something that happened in the parking lot at work later that evening, after we had returned from our celebration.

The day had been long. No sleep the night before, up at 4 A.M., a series of emotional highs in rapid succession, and now we were back where the day had started about 18 hours before. A final goodnight. People took off in their cars, headed for home to share the day's events with their families—all except Hub and me. I could sense he wanted to share something, perhaps one final recap of the day's events.

He was very quiet; perhaps he had no energy left. I started to talk about tomorrow and all the things that still needed doing. Slowly, he reached for my hand and touched it. There were tears of joy in his eyes as he started to speak. "Thanks for making all this possible. Our team has lived many lows in the past two years, but the way we have come to care about each other has made it all worthwhile. Today is a memory I will treasure forever."

Driving home that evening, I couldn't stop thinking of Hub and of what had just happened. A powerful emotional bond had grown between us and among all the members of the Copper Shop team. I thought back to the management appreciation breakfast held in March 1986, by the production specialists in the PIC insulating shop. That breakfast had been their way of saying thanks to the managers and me for changing, for caring, and for giving them a renewed sense of pride in their work. I thought about the focus team session Marilyn had conducted the following August to learn why the production specialists cared enough to give us such unprecedented recognition. And drifting through and around and among these memories, like a strain of music that stays with you for hours, were Jerry Butler's words of insight and understanding: "Care makes it work."

We don't know who we are until we see what we can do.
—Martha Grimes

— ✧ —

The only things worth learning are the things you learn after you know it all.
—Harry Truman

— ✧ —

Faced with the choice between changing one's mind and proving that there is no need to do so, almost everyone gets busy on the proof.
—John Kenneth Galbraith

4 LEARNING FROM THE WALL

SOME EXPERIENCE REQUIRED

With remarkably good timing, Lew's story from the Copper Shop had provided an ideal, double-blind case study, validating the cultural study findings and showing parallels to all the key details. Earlier the study team had uncovered, beyond any doubt, the fundamental structure of the American quality archetype; from the Copper Shop there was now a rich description of the emotions and experiences that surround it and the consequences and behaviors that stem from it in daily life. Okay, now what? The cultural study and the Copper Shop investigation were finished. This new learning was too precious to let vanish like yesterday's interoffice mail. The study team wanted to learn more; they wanted to continue validating the findings and disseminate this information as widely as possible, in a way that would produce some tangible benefit. What did the archetype really mean, anyway?

The team's discussions were personally challenging, almost confrontational. They would assault each other with new ideas, follow what appeared to be logical paths, then someone would toss in a thought that came straight from outer space, and they would begin second-guessing all their previous assumptions. Could other people merely be told about the archetype? With only an academic, qualitative description, would others find the information useful?

The team finally decided that an oral presentation or written report wouldn't be sufficiently effective. Instead, they wanted an experience or event that would recreate the archetype, something that would force people to discover it by feeling it before they had a chance to think about it. Imprint studies have shown that the deepest learning takes place when people experience something emotionally, so the team decided to give people an appropriate emotional experience, something that would show the archetype's process in action. They knew this was vital for the learning to have a long-lasting impact.

MARILYN LEARNS A PROCESS

Once in the late 1960s, a search for student-directed learning took me to New York to look at a federally funded education endeavor called the Education Development Project; it was based on Jerome Bruner and Jean Piaget's theories of learning. I had read of their work distinguishing between product and process learning. When educators talk about a product, they often mean a product of rote memorization; this includes things like multiplication tables, the five major exports of Bolivia, and other discrete packages of information that traditional education tries to wrap up and hand to students as attractively as possible. In the same context, product can also refer to educational approaches that encourage students to copy an existing model of something. On the other hand, process refers to the active process of learning.

The people running the Education Development Project had set up a learning center in a nondescript loft on Canal Street. What I found inside was, at first glance, chaos: heaps of what I can only call "stuff" scattered over hundreds of square feet; the stuff included scraps of lumber, power tools, hand tools, film and audio/visual equipment, books, children's toys, a wide variety of art supplies, a sand box, and many other items. I knew already, mostly from watching children, that this was the perfect environment for early learning; but its mission here was to create a learning laboratory for teachers, most of whom worked in the New York City public school system.

The stuff strewn around the loft's considerable open space was there for people to explore, to probe for possible discoveries. The attitude in the center was relentlessly open-minded; any beginning was a good beginning, and all pathways were equally correct. But the primary learning the center hoped to achieve was the difficult shift in focus from product to process. Teachers came there as students to gain true understanding of the learning process by discovering it naturally.

The workshop offered absolutely no focused, authoritative direction; this was the central idea, to get people to learn by doing, by discovery. The basic activity that we all seemed to fall into was making something, anything, out of the stuff scattered around the loft. We connected, disconnected, appliquéd, nailed, screwed, and cut and pasted. We students would become tremendously energized and excited in the workshop, especially when using materials and tools that were beyond our normal experience; electric saws and drills were particular favorites. We usually built familiar things, rarely inventing anything new. But in either case, this activity was totally end-product focused—our natural tendency.

The center's facilitators tried to push us in another direction. They subtly nudged us to explore, question, and observe the materials and processes we used to create our products. We students, all working to construct something, would occasionally let go of our product focus to observe our process; but in general we were strongly focused on whatever we were making. Our observation of process was just a moment's diversion. Most of the time, the teachers chatted delightedly with each other about how they couldn't wait to "teach" their public school students the steps they'd taken to construct their current masterpieces. We were all stuck in the product focus. As I watched these fine imaginative teachers, nonetheless stuck, I began to see the difference.

One evening, the facilitator suggested I play in the sandbox; it was an unusual sandbox, containing sand of different colors and grains of different sizes. I also found a number of sifters with different size filters. I started to play rather casually, first making random designs, then seeing how the sand behaved in motion, and then noticing behaviors that were common by color but different by size. I began to wonder why.

Looking for fast answers, and having no patience to discover them myself, I asked the facilitator; he responded with more questions. This process of play with the sand, ask a question, get an intriguing response, and play some more continued until the center closed for the night; by then I was hooked. In the space of a few hours, I had dug deeply into physics and chemistry, two subjects that had never before held much interest for me. I had exhausted all their books that contained even a shred of relevant information and was ready to hit the New York Public Library as soon as it opened … and then it hit me. I was engaged in a learning process, not just creating a product. I was learning how to learn. What I didn't know then, but know now, is that I was also learning about quality.

In that educational laboratory, the difference between product and process differentiated teaching from learning. Product-focused teaching is intrinsically limited, and its benefits are often short-lived because it is nontransferable. Learning is unbounded, is utterly free from restrictions, and often lasts a lifetime. For too long, American education has emphasized product-focused teaching and has ignored learning about the learning process itself. It's very much the same in American business. Our business environments emphasize end results, with little respect for learning. Even in the quality approaches that emphasize a process rather than a product focus, the process is task-oriented, not learning-driven. Perhaps this needed shift could begin by one's seeing the parallels between business problem solving and playing in that sandbox.

THE WORKSHOP

To provide the deepest and most valuable learning about the quality archetype, the Network Systems quality team wanted to offer people an active learning experience. They gave them one: a three-and-one-half day workshop under the title "Building Total Commitment to Quality."

The workshop was designed as a collection of activities, presentations, and discussions that enabled people to discover the archetype by reexperiencing it.

Unlike the cultural study process, it wasn't necessary for each participant to revisit his or her own unique first experience. Instead, the team's intention was to create an environment in which people would naturally undergo the same basic set of circumstances, stresses, and emotional responses that accompanied that archetypal first experience. Because the cultural study had shown how the American archetype functions, the team already knew what to expect and felt that the archetype's telltale signature would be readily visible in the participants' behavior.

The workshop followed the same three-phase process as the archetype: Crisis and Failure, Support, and Celebration. It actually doesn't take much to put Americans into a state of crisis, an emotional experience that's just as painful and embarrassing as failing a homework assignment or losing a big client.

A FAILURE FOR PHASE I:
PREPARING TO DO IT RIGHT

Hotel conference rooms are similar all over the country, and most of the workshop participants had been in them before; off-site events and meetings are fairly common at AT&T. Upon entering the room, the participants in the group can immediately connect with familiar things. The room smells like air that's passed endlessly through air-conditioning systems, carpet that's vacuumed every day but still holds a memory of cigarettes, and coffee that's well on its way to being burned. Stackable hotel chairs line some of the walls. Folding hotel tables, skirted with pastel tablecloths, bear fresh fruit, a few dessert items, coffee, plates, cups, and napkins. A TV monitor and VCR on a cart stand in one corner. Everything is bathed in fluorescent light. Dressed in casual clothes, the participants stroll in, absorbing this simple and familiar setting. They smile, joke, try to act as casual as their clothes, and conceal their undercurrent of tension.

The tension comes from things that aren't so familiar. The chairs aren't arranged for any kind of conventional seating. A video camera on a tripod stands near the middle of one wall. And strangest of all, there's something running across the middle of the room, about 4 feet above the carpet, that appears to be some sort of temporary wall. On one side of it, the floor is covered by spongy, vinyl-covered workout mats.

The workshop facilitators greet people as they come in: "Hi, how are you, thanks for coming." The participants try to lose their apprehension in coffee and shop talk; in a few minutes, the facilitators get started.

Following the pattern of the archetype, the workshop's facilitators function initially as Lawgivers. They create a sense of urgency by immediately handing the participants a project and laying down a very simple, very clear set of rules: "That wall you see going across the room? Well, pretend it's reinforced concrete, two feet thick; and not only that, it's also electrified with 1,000 volts. All of you are going to start together on one side of the wall. Your job is to get everyone to the other side, over the wall. You can't go under it or around it, and you can't use any furniture. Anyone who touches the wall dies. We'll be videotaping this exercise, and you have 45 minutes to get everyone over alive. Okay, go ahead."

The participants don't like this one bit. It's ridiculous, silly, and clearly a waste of time. They came here to learn about quality, so what was this wall business? All of them feel uncomfortable, thrown off balance; they're trapped by a stupid, improvised wall, for God's sake! But freedom lies on the other side. Time is their enemy, and if they want to escape, they must do it in 45 minutes.

Silence. This is that rare moment when thoughts are nebulous, formulating, subduing people's actions. They look around, somewhat blankly; several approach the wall, examining it from all angles. Others form small groups at the outer edges. A few stand alone, chin cupped in hand, pondering. Suddenly, the questions erupt, irrational, emotional, erratic, and in complete denial of the rules they've just heard: "Can we go around the wall? Can we use that table or those chairs over there? Can we short out the power so the wall isn't electrified any more? Do we really have to get everyone over? What's the trick? Has anyone ever done this?" These and many other questions pour from the participants in a flood.

The facilitators are very understanding. They repeat the rules and even restart the clock: "You have 45 minutes."

One or two individuals begin to give serious thought to solving the problem by actually observing the rules. Others stand idling at the perimeter, frozen, hesitating, waiting for someone to rescue them from their miserable sense of helplessness. Why don't the facilitators just tell them what to do? Some participants form small subgroups, still plotting a rule-defying intrigue to beat the system.

The noise level from separate conversations rises. Some members of the group suddenly jump into action, heading daringly toward the electrified wall. They seek adventure and glory.

Others continue to linger about the edges, unable to commit themselves, adopting instead a skeptical, judgmental attitude and disdainful posture. Someone cracks a joke, everyone laughs, and many express doubts about the chance that this silly task could ever be successfully completed.

Now people are all talking at once. Rapidly, idea after idea is launched and immediately drowned in the din. Those who shout the loudest ultimately command the attention of the group and emerge as leaders. A facilitator speaks up: "You have 35 minutes."

"Build a pyramid ... let's use our bodies as a human ladder so someone can climb up and jump over," is the first clear suggestion from one of the strong-willed men. Someone drops onto hands and knees; others do the same. "Get closer, we need three strong men on the bottom! Okay, get two others on top!" This flurry of activity draws the entire group toward the wall. For the first time, they all have a sense of unity and shared purpose. Everyone has the singular hope that this will be the master stroke, an obvious, quick and easy triumph, which is what they all want.

The pyramid collapses. A facilitator speaks up: "You have 28 minutes."

Although dreams are dashed, the idea is abandoned, and failure is acknowledged, some interesting things begin to happen. The participants acquire a greater respect for the challenge and, surprisingly, a greater regard for the value of each member of the group—a shadow of cohesion and team spirit makes a tentative debut. Now, still divided into several subgroups, they begin again. They speak more quietly, somewhat subdued by their initial failure.

A facilitator speaks up: "You have 25 minutes."

"Throw someone over!" Panic strikes hard and fast when they realize how little time remains and still not a single person is on the other side. Without discussion, several men decide to pick someone up and toss him over the wall. They eye their victim and motion for him to join them: "Come over here and lie down next to the wall." His face shows panic and real fear. "We're going to lift you up and toss you over."

They remove eye glasses and jewelry, the barest token of regard for the frightened guinea pig's personal safety. This is now an act of survival for the whole group, and time is working against them. Their idea sounds good, and if it doesn't work, well, they'll find a better way next time. "Okay, ready, lift!" Using brute strength and sheer determination they begin lifting, hopelessly uncoordinated.

Clumsily, they raise the body dangerously close to the wall. Not a single thought has been given to death by electrocution. No trial runs have been made. No one has suggested they experiment first, maybe in another spot, farther from the wall.

The victim, rigid with fear, is slowly raised to the top of the wall. The lifters' voices groan, muscles strain, and beads of sweat appear on their faces. "What next?" someone shouts. The reply: "On the count of three, throw him over the wall." The victim's protests go unanswered. In a tense, strained assembly, drawn together and intent on the action, the group is deaf to the man's plea.

"One, two, three." Within seconds, it's over. He hits the mats dead on, and guess what …"I did it!" he shouts. The man is a symbol of the power of determination! The hapless guinea pig has been transformed into a hero—he did his part, bravely faced death and triumphed, not just for himself but for the group.

Rousing cheers celebrate the group's victory. People are smiling and happy, many hands slapping high-fives, like football players after a touchdown. The gallant men who did the lifting are heroes. They bravely challenged the wall of death and won. Energized and confident, flushed with success and obvious mastery, the group now has more than enough volunteers. All they have to do is repeat this first performance a few times to complete the task. No one sees any need to explore other possibilities or to improve on the method.

A facilitator speaks: "You have 20 minutes."

The participants spring into action. At a frantic pace, several more volunteers are safely hurled over the wall; the remaining members stand passively about, mere spectators at this point. The facilitators watch this busy activity and see a number of participants touch the wall. These individuals either don't realize they've touched the wall or do and are trying to ignore it. As yet, the facilitators don't say anything.

With continued and easy success, the group begins to lose interest in the task. "At this rate, we'll beat the clock," someone says. In the space of a few minutes, the team becomes complacent and careless.

ZAP!

"You're dead!" shouts a facilitator, standing like a tennis referee at a monitoring post. A member of the group has touched the wall. Everyone stops. The mood becomes somber, silent, mourning. Almost instantly, someone challenges the call. "He didn't touch it." Someone else offers an excuse that is instantly supported by the entire group. They look at each other and smile, and the pace of activity increases once again. Within minutes, they're eagerly and recklessly back into their work.

A facilitator speaks: "You have 10 minutes."

As the work continues, the flow of ideas begins to slow down. What little experimentation there is takes place center stage, right next to the wall, clearly the most dangerous place. Everyone continues to rely on physical strength long after they should have looked for other methods. Excitement mounts as more people make it over. Encouraging shouts of "We can do it!" burst intermittently from people on both sides. Victory is surely within their grasp.

The presence of a new challenge occurs to the participants. Most of the lighter, smaller people have already been put over the wall, leaving mostly the taller, heavier men on the original side. How will they get over? Are they up to the task? Panic begins to appear and spread. All activity comes to a halt. They huddle, some people suggest several new ideas, and within minutes the team succeeds in lifting a big man over the wall. They all rejoice with unbelievable glee, but their celebration doesn't last long.

A cry of "You're dead" cuts through the air, called out by one of the referees. A hush falls over the group. Someone shouts, "We have ten minutes!" and their mission continues with an even stronger sense of urgency.

Now, with greater and greater frequency, the call of "You're dead" rings out in the room. No one monitors their movements to prevent contact with the wall, and still no one has suggested that experimentation be done away from the wall. Instead of watching for defects in their approach and execution, the participants watch the facilitators. When the participants know the wall has been touched, but the facilitators don't enforce the rules, the activity continues with even greater abandon (maybe they can get away with it). When the facilitators point out that the wall seems to be swaying a bit and ask if any of them touched it, the participants offer a remarkably creative alibi: the wind. "The wind moved the wall," they say, "not us."

"Oh, okay. You have five minutes."

No time to lose! There are still three people remaining on the original side. The oppressive, awful, horrifying, unanswered question, which everyone has felt from the beginning, demands to be heard at last. "How will we get the last person over?"

Succumbing to the need for comic relief, people joke about the situation, their smiles too broad, voices brittle with tension. "Well, it's been nice knowing you." Laughter surrounds a frozen pocket of space, where standing alone are the stranded strong men who had provided the brawn, once heroic and full of pride, now doomed.

"You have three minutes."

Quickly, the people pour out ideas. Many are criticized and rejected, others are ignored completely, and hardly any are actually tried. If an idea is tried and success is not immediate, no one considers attempts to improve or refine it. No one adds to another's thought in the hope that ultimately they'll find a workable solution. The group struggles feverishly and manages to get only two more over the wall.

"You have one minute."

Someone shouts a command. "You, you, and you line up at the wall. We're going to lift him up and over with our hands. Come on, let's go." Someone challenges, "Wait, he's too heavy, we won't be able to hold him up." The new leader replies, "Quick get another guy over here to help. Two of you on that end hold his arm and shoulder, he'll raise his leg horizontal to the wall, we grab it, hold him out away from the wall and lift him over."

Someone expresses doubt, but the new leader is unstoppable. "On the count of three, we lift; one, two, three ... ZAP! ZAP! Two more people perish at the wall as they try to save the last hero.

Silence. Dismal and defeated, each participant collapses into a desperate, inward search for meaning. Some appear sullen and isolated, others dismiss the whole thing as a game. Someone moans, "I can't believe this happened."

"WHAT HAPPENED?"

Crisis and Failure. All the participants feel it as profoundly as if they'd just erased some vital files on a computer, lost an essential market to a competitor, or missed the final deadline for launching a new product. They are all deeply in Phase I of the American quality archetype. A few people act like they were unaffected, but most are clearly, openly embarrassed. They're in the middle of a big mess, and they feel it very deeply.

The next step in the workshop was a simple debriefing. We pull the chairs into a circle, sit down, and begin to talk about the experience. The facilitators remind everyone that this was only a game, with no right or wrong way to do it. They then encourage the participants to talk about what happened and how they feel.

The conversation is very slow to start. No one feels comfortable talking about it. Finally, someone speaks up. "Well, it went okay," and another, "Yeah, we did it except for the last person. If we'd had more time we'd have picked him up and lifted him over the wall." This is fascinating to observe. There's a gradual shift in the perception of what happened, from utter failure to reasonable success. One by one, they confirm and amplify each other's belief that they did a fine job and, by gosh, they feel pretty darn good about it.

"How many died?" we ask. They reply, "Well, maybe three or four died." We insist on sticking to the facts. "We saw seven or eight die." They refuse to accept the facts: "Seven or eight! No way!" Another dodges behind the planned inconsistency in the facilitators' calls: "You called some dead and not others. How can we know when you're not consistent?"

Round and round the circle come excuse after excuse, defensive comments, an unwillingness to disclose the truth of their behavior. When asked if the patterns of behavior in the exercise show any similarity to those on the job, some feel that they do and make a few superficial comments. "Yes, I usually try to be helpful," from someone who volunteered to be one of the first over the wall. "When I'm on a project, I pitch in from the start," from the strong-armed fellow who lifted the first and last few over. Others deny that their work is anything like the exercise.

The comments become more general. "We have similar problems on the job, when we're trying to solve a problem with a team. Everyone talks, no one listens. We get into an idea and run with it." With this comment, several others remember specific incidents and tell stories about a mess they were in at one time or another. In some cases the stories are still fresh enough to hurt. In older, half-forgotten stories, the poignancy and distance bring laughter.

This discussion continues for 45 minutes. Several times, people ask for the solution; some have an insatiable appetite to know how they were really supposed to do it. Toward the end of the session we ask, "Is this the way you approach quality on the job? Do you cover up mistakes?" This question strips them naked. Dead silence falls and the tension in the room shoots way up. People fidget uncomfortably in their seats. Embarrassment bores a hole deep into each person.

For a few deadly minutes, an icy silence envelops the group. And then, one by one they attack us, letting us know in no uncertain terms that we, the facilitators, are at fault, not they. Their words are harsh, their tone unkind and accusing.

"How can you expect quality when you don't explain the rules clearly? When you aren't consistent?" from one. Another throws out: "Quality? I don't think it's possible to do this exercise. There is no solution."

The session comes to an unpleasant end. We announce our next starting time and disperse. Inevitably, a few people stay in the room, determined to find the trick. Some actually remain behind for hours, staring at the wall, experimenting, unwilling to let it go.

The first phase of the workshop, and almost the first phase of the archetype, is past, and the facilitators head for their rooms. They're not troubled by the participants' lack of honesty or their unwillingness to probe deeply for insight. This is Phase I: Crisis and Failure. The participants' behavior here is precisely the behavior of individuals and teams when they begin a quality effort, although it's not so obvious to the group.

They failed to meet the requirements: getting everyone over the wall alive in 45 minutes. To make matters worse, they were more imaginative about finding ways to hide their mistakes than ways to confront them. In this phase, it is virtually impossible to create a learning situation; all participants are so confident in their ability to perform the task that they are closed to learning anything. This group is no different from the many that preceded it and the many that will follow. More than 35 groups, about 600 people, went through this exercise, and every group behaved exactly the same way—Phase I of the archetype was validated every time.

As much as Americans would like to believe that their projects are carefully planned and precisely executed, the facts show that American projects and their associated project teams go through Phase I. Sometimes they never get out of it.

THE PATTERNS

The study team expected the archetype to predict people's behavior during this exercise, at least to some extent. They expected that people wouldn't complete it without some mistakes and embarrassment, but they were unprepared for the consistency of the results and the accuracy of the predictions. The archetype served as a virtual crystal ball. It was amazing: every group behaved in exactly the same way. Figure 4.1 shows observations from all the sessions, grouped under headings that characterize the behavior.

Every group is sloppy. Rather than plan, the group's first choice is to take action. Many touch the wall but continue in the game anyway. When the facilitators point out that the wall is moving, every group invents "the wind." The wind did it. No one serves as a spotter to watch the others, no one monitors the quality of the

BUILDING TOTAL COMMITMENT TO QUALITY WORKSHOP WALL EXERCISE: PATTERNS OF BEHAVIOR

HOW WE DEAL WITH GETTING THE TASK DONE
- We celebrate getting the job done regardless of the quality
- We're sloppy and careless

HOW WE TREAT RULES
- The rules are never clear
- We don't really understand the rules
- We break the rules
- We test the rules
- We try to change the rules

HOW WE DEAL WITH MEN AND WOMEN
- Men form at the center
- Women form at the periphery
- We have to be center stage to get attention

HOW WE TREAT PEOPLE
- We throw people over the wall
- We practice at the wall where risk is highest
- We try macho strength, brute force, or our backs to get the jobs done
- We move to action ASAP
- We don't listen to one another
- Some groups sabotage one another

HOW WE TREAT IDEAS
- We shout out ideas but never test them
- We'll get stuck on one idea and continue to use it even after it stops being useful
- We don't understand ideas until we try them
- We don't plan

HOW WE DEAL WITH LEADERSHIP
- Persons willing to use themselves come closest to being leaders
- We don't allow leaders to emerge
- No shared sense of purpose emerges

HOW WE FEEL
- Awkward and afraid
- Embarrassed
- Some cover it up with humor or bravado

HOW WE DEAL WITH TRUTH, REALITY, HONESTY
- Our creativity is for alibis, not solutions
- We collaborate to cover up defects
- Every group invents "the wind"
- When the exercise is over, we continue to defend and cover up reality
- We identify the most difficult problem (last person) and ignore it
- When the exercise is over, we have difficulty seeing and acknowledging what really happened

Figure 4.1

operation or guards against defects or mistakes. No one worries about safety. In some groups, people watch the facilitator when they know the wall was touched. If the rules aren't enforced and they're allowed to get away with it, they work with greater abandon. If a facilitator tells a participant he or she is dead, the person usually honors the call and drops out, but often other people try to pull the casualty back into the exercise, and the person may actually join in again.

As the exercise continues, the flow of ideas always slows down. The ideas generated closest to "center stage," the area offering the greatest visibility, are usually the ones people follow. Most ideas are ignored or rejected, and no one suggests the approach of cumulative learning, assembling the best elements from multiple ideas. Ideas are always tested right next to the wall, with no concern that the effort may cause them to touch it. There's never any attempt to practice, to try an idea in a safe place before executing it for real.

During the first few minutes of the exercise, every group identifies the most difficult task in the exercise—getting the last person over—and then ignores it. Only when two or three people are left does the group begin to seriously think about how to get them, and especially the last person, over the wall. No group successfully completed the exercise on the first attempt. Doing it right the first time is not our natural American quality archetype approach.

RECOVERY

The next morning, the group assembles again, in the same room but without its wall and workout mats. Now the tables bear cereal, danish, muffins, juice, more fresh fruit, and "younger" coffee. Together, the participants and facilitators watch the video recorded during the previous exercise and get into a much deeper group discussion.

The participants learn the discrepancy between their beliefs about their behavior and their actual behavior, especially regarding quality. Like it or not, they can't deny what they see on the screen. The video forces them to accept the reality of what they did during the exercise. Most are deeply embarrassed and cope by laughing and poking fun at themselves and each other. Some want to try it again right away, because they believe they've discovered the trick of how to do it.

People demonstrate superb creativity during Phase I, and this behavior is clearly evident in the workshop. The study team called it "alibi creativity."[1] As shown in Figure 4.2, the workshop participants are creative in finding reasons to explain why the wall is moving and in defying or outmaneuvering the rules.

During the first exercise debriefing, they're creative in identifying what they feel is the truth about their behavior: they are innocent, and all faults rest with the facilitators, the Lawgivers. The facil-

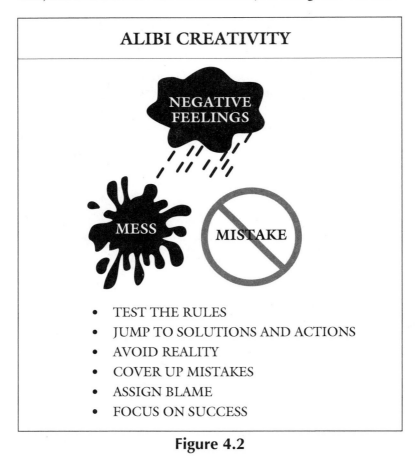

ALIBI CREATIVITY

NEGATIVE FEELINGS

MESS

MISTAKE

- TEST THE RULES
- JUMP TO SOLUTIONS AND ACTIONS
- AVOID REALITY
- COVER UP MISTAKES
- ASSIGN BLAME
- FOCUS ON SUCCESS

Figure 4.2

[1]Rapaille, *Creative Communication.*

itators are blamed either for not telling them when they touched the wall or for behaving inconsistently. Note that the rules didn't include a job description of the facilitators as "wall police." Few admitted freely that they touched the wall and did not reveal the failure they knew about.

This Phase I behavior is characteristic of all Americans. As much as they'd love to act differently when problems occur, they don't. Efforts to change this denial behavior are futile. In fact, trying to change this behavior is often the very thing that dooms American business efforts and foils American attempts at quality improvement. No amount of investment in "should" programs, which try to get people to do things the way they should the first time, the way that makes logical sense, will ever guarantee the desired improvement. These programs all fit under the umbrella "ways to help us do it right the first time" and therefore deny the existence of Phase I of the archetype structure.

It's a mistake to see this as some sort of disease or weakness, something that must be battled, like cancer. It's far better to look at the archetype as a valuable, unconscious American pattern of how to do things right. The initial failure provides emotional fuel. In a supportive environment, it inflates people's desire to correct their mistakes and do the job right. It's important for Americans to go through a stage of initial failure because they don't truly understand the rules or learn the process until they act, until they make the task concrete by doing something.

The most immediately applicable idea from the archetype is this: because Americans are so driven to do, they should do prototypes—do fast prototypes. Don't try to make it right the first time, just try to get it made, period. Get it produced, done, fast. Once the prototype is finished in some form, the makers should encourage challenges. The question, "What's wrong with it?" now becomes tremendously useful and can help the makers and other concerned parties see what aspects of the prototype need improvement. Don't think this sort of process is limited to operations that stamp out widgets. Any product, service, or idea can be created as a prototype, even if it's a simple improvement to an existing product or process (see Chapter 5 for more information on prototypes).

When there's no penalty for a first attempt that's less than perfect, and not expected to be right the first time, people's egos are not so vulnerable, and the challenges and criticism such first attempts draw can be extremely helpful. First attempts are still personal creations and not totally free of ego, but any guilt associated with not doing it right is far less toxic and paralyzing.

PHASE II: LEARNING TO DO IT RIGHT

With the end of the video-supported discussion of the wall exercise, Phase I is over. The workshop participants have, to some extent, put their crisis and failure behind them.

As the workshop continues, the facilitators, initially Lawgivers, now become Coaches and Mentors (not a recommended switch in a real business situation). The participants work through a series of modules that provide activities, quality examples from AT&T and other companies, and stories from the lives of particular individuals and teams. During these modules, the participants experience creativity, visioning, empowerment, support, and caring.

In one exercise, participants describe what it feels like to care for someone and for someone to care about them; Figures 4.3 and 4.4 show a composite of the actual responses. One of the more interesting patterns that emerged is that participants repeatedly connected feeling cared for with freedom. Freedom and feeling free are powerful personal attributes, particularly for Americans. Stories of liberty and escaping oppression are common among stories of early American history.

After each module in the workshop, the group evaluates its experience along eight criteria. Immediately after every evaluation, the facilitators calculate the mean for the class and display the results. During the course of the workshop, this lets everyone see the curves representing the group's impressions about each module.

Some reactions are consistent across all groups. The curves for participant creativity and leadership are always in complete opposition—the more the leaders (facilitators) are felt to be in control, lecturing or otherwise giving information to the group, the less the participants feel they're using their creativity. When

WHAT DO YOU DO WHEN YOU CARE FOR SOMEONE?

- REACH OUT
- EMBRACE, HUG
- BE THERE PHYSICALLY
- UNDERSTAND
- COMPASSION
- PROVIDE DIRECTION
- ACCEPTANCE—AS IS
- SHARE DREAMS
- BE CREATIVE WITH THEM
- SHARE
- BE RECEPTIVE—OPEN COMMUNICATIONS
- ACKNOWLEDGE ACCOMPLISHMENTS
- DO NOT FEAR BEING WRONG
- CARE EVEN UNDER EXTREME STRESS
- BE HONEST
- TRUST
- ENCOURAGE
- SUFFER WITH

- BE A CHEERLEADER
- TELL YOU CARE
- RELATE FEELINGS
- RESPECT THEM
- BE POSITIVE
- SHARE NEGATIVES
- GIVE THEM A SMILE
- GIVE THEM TIME
- GIVE THEM RECOGNITION
- YOU LISTEN
- YOU TOUCH
- PERSONAL SACRIFICE
- PROTECT
- SUPPORT
- REASSURE
- REWARD
- CELEBRATE WITH
- LET GO—GIVE/DESIGNATE RESPONSIBILITY

Figure 4.3

the participants feel their creativity is high, the leadership quality is rated low. The message from this response is clear: if you want people to feel creative, you must give up some of your leadership position to them.

WHAT DOES IT FEEL LIKE TO BE CARED ABOUT?

- SENSE OF PURPOSE
- VALUABLE
- ALIVE
- RESPONSIVE
- TOUCHED
- ACCOUNTABLE
- POSITIVE OUTLOOK
- GRATEFUL
- EXHILARATED
- RISK-TAKING
- FAMILY
- RESPECT
- FREE
- IMPORTANT
- GOOD
- SAFE
- WILLING TO GIVE BACK
- MORE OPEN TO EXPRESS YOURSELF

- ADVENTUROUS
- NEEDED
- ELEVATES SELF-ESTEEM
- BOOSTS MORALE
- WARM AND FUZZY
- SPECIAL
- PROUD
- NICE
- LOVED
- INVOLVED
- WORTHY
- BLESSED
- YOU MAKE A DIFFERENCE
- FULFILLED
- HAPPY
- ENVIED
- WILLING TO WORK HARDER
- LOYAL

Figure 4.4

If you're ever in charge of a project, you are almost certain to reach the point when mistakes begin to occur. In your desire to correct the situation, you may then find yourself inadvertently taking control: you jump in, give orders, reorganize, and generally crack the whip. Consider another approach: help those who have made the mistakes express their embarrassment and guilt. Provide coaching and mentoring to those who really need to correct the situation. This will help them convert their negative energy to positive energy, to fix whatever needs fixing. If you can't give this kind of support, because you are emotionally charged by the consequences of potential failure, find someone else who can, someone who can act as mentor and coach for the individual, group or team.

Another important lesson from the workshop, not directly related to the archetype, is the importance of play and of having fun together. Play was built into the workshop agenda and purposely given the same degree of importance as any of the other material, not as some optional after-hours activity; the study team came to feel that this is a critical issue for business. Managers are encouraged to begin building times for fun and play into the American workday. Playing, particularly physical playing, is perhaps the best way to rebuild people's energy, while creating closer bonds among team members.

As a play activity, the workshop group played wallyball. This fiendish athletic invention amounts to volleyball on a racquetball court. The facilitators also tampered with the rules to ensure maximum participation, exercise, and fun. Before leaving for the courts, the group agreed on the objectives and rules. They were posted visibly, as with the other ideas from previous modules. When the game was over, the participants evaluated it similarly to the other modules. One immediate consequence of this activity was that the dynamics of the group improved dramatically. As well, this was the first element of the workshop that most participants transferred to their workplace: the regular inclusion of play. It was consistently the second-highest-rated module.

A MODERN AMERICAN FABLE

After the wallyball game, the group has dinner and then watches a movie: *The Karate Kid*.[2] While by no means an immortal work of cinematic art, this film has a strong, well-crafted story and offers a wonderfully clear example of the American archetype in action, including excellent portrayals of the mentoring and coaching roles.

[2]William Idol of William Idol & Assoc., Ltd. is credited with the analysis of *The Karate Kid*, Columbia Pictures Industries, 1984.

For those who haven't seen the film, it's the story of Daniel Lorenzo, a displaced high-school student, just moved from New Jersey to California. Daniel is basically a nice, intelligent kid; but he has something of a chip on his shoulder. As the story develops, his pride continues to be his worst enemy.

One day, in his new neighborhood, Daniel meets the Girl of his Dreams. The chemistry between them seems wonderful; but unfortunately, this young woman happens to be the estranged girlfriend of a major football star and martial arts expert at Daniel's high school. Inevitably, Daniel and this young man come to blows, and Daniel is definitely somewhat worse for wear. Things look grim, until the appearance of a kindly Japanese man, Mr. Miyagi, the custodian of Daniel's apartment complex. At a particularly dangerous moment, Miyagi comes to Daniel's rescue and employs superior martial arts to save him from a dangerous beating at the hands of the high school bullies.

Daniel and Miyagi become close friends, and Miyagi teaches him that there's more to life than revenge and physical prowess; but as events progress, even the peaceful Miyagi realizes that Daniel must learn to defend himself. He agrees to teach him karate. Daniel works hard and follows Miyagi's teachings. At the film's climax, he competes with his arch-nemesis at a local karate tournament. He wins first prize and the admiration of his girlfriend, mother, and teacher.

All through this movie are examples of archetypal behavior and many lessons for today's managers. The first time Daniel is beaten up by the local black-belt, who was provoked by Daniel's attentions to his ex-girlfriend, Daniel lies hurt and embarrassed in the sand—his first crisis. The girl, Allie, tries to comfort him, but he rejects her, wanting only to be left alone. This is very common behavior. When people are hurt and embarrassed, they're not very rational. It's a bad idea to pay strict attention to what they say. Under these circumstances, you may think it's best to leave the person alone. It's certainly easier to leave the person alone; it's very uncomfortable to deal with other people's emotions, especially negative ones. But it's better to stay close to the person and listen. Don't attack his or her behavior, don't demand explanations, just be there when he or she is finally ready for you.

Later in the film, out of anger and frustration with his worsening situation, Daniel throws his bike into a dumpster. Both his mother and Miyagi witness this event. Neither of them knows what has provoked the action and can only draw conclusions from his behavior. His mom responds with a frantic barrage of questions: What's wrong? Why are you acting this way? Daniel counterattacks his mother for bringing him to California in the first place. If you want to mentor someone in Daniel's state, you cannot take their attacks personally.

Another thing to note here is the difference in energy between mom and Miyagi. When this argument takes place, Miyagi watches silently from the background. Daniel's mom uses her energy as a frontal assault to pressure him, trying to get him to tell her what happened. Miyagi uses no energy at all but soon begins to act in ways that draw Daniel toward him. He rescues Daniel's bike from the dumpster, repairs it, and leaves it so Daniel will see it the next day. Miyagi begins to build a relationship with Daniel and expends very little energy doing it. He creates a situation and lets Daniel use his own energy to approach him. This is a basic principle for developing people. Don't try to use your energy to fix someone else's life. Instead, create situations in which people can learn to fix their own problems.

Daniel defines his problem as not knowing karate. Without it, he can't defend himself; of course, that's not his problem. He's not ready to see the truth—that he has played a part in provoking the attacks (he has a wiseguy's chip on his shoulder). Daniel is firm about learning karate, and if he doesn't think you can teach him, he won't let you mentor him. The unspoken message is: "I won't let you help me unless you deliver what I think I need."

Later, Daniel finds Miyagi pruning a bonsai tree, and Miyagi suggests that Daniel prune one too. When Daniel protests that he doesn't know how, Miyagi tells him to close his eyes and picture a perfect tree. When he opens his eyes, Miyagi tells him to make the bonsai look like his imaginary picture. Miyagi is coaching Daniel to have and use impossible dreams. The lesson here is to respect and use the power of visualization and imagination. Pictures are more powerful than words. The visions people create for themselves, their teams, and their organizations can be far more powerful if they're converted to symbolic pictures, particularly when such visions reflect impossible dreams. When things go wrong in Phase I, trusting the image of a dream can pull people through.

Miyagi continues to build a relationship with Daniel. Later in the movie, he rescues him from another beating, attacking the bullies after Daniel is struck unconscious. Later still, he gives Daniel a bonsai tree and makes him a costume for a Halloween party. He has done much for Daniel and so far has made no demands. None. Soon he will make demands, but he understands, as many of us don't, the need to build an emotional foundation before expecting Daniel to meet those demands. In our organizations today, bosses often automatically make demands, simply because they're in positions of power. But there is usually no emotional basis for anyone to want to fulfill these demands. When you understand emotional reality, you realize there are some things you invest in before you make demands of someone else.

Miyagi lets Daniel know he rescued him using karate. When Daniel asks Miyagi to teach him, Miyagi asks why. Daniel replies that he wants to be able to fight because he's always getting beaten up. Again, that's not the problem, but rather the symptom. The problem is the way Daniel acts with people. He's cocky, he's a wiseguy, and his behavior provokes the beatings. People tend to focus on symptoms because that's where the pain is. People generally move to alleviate symptoms, not causes. Initially, it's much more difficult to go after the cause and, often, the cause is not located close to the symptom. Too many analyses of root cause focus on the symptoms and consequently never uncover the real cause.

Daniel asks Miyagi to go to the karate school in town where the bullies train to make a deal to keep them from beating him up. Here, Daniel has asked Miyagi to fix his life, and Miyagi responds by testing the limit of Daniel's capabilities and self-confidence. You can't believe what Daniel says because he's feeling weak right now. He'll tell you that he needs your help, even when he doesn't. Where is the real boundary between what Daniel can and can't do, and where is the perceived boundary? When Miyagi refuses, Daniel threatens to break off their relationship and leave. The boundary now becomes clear.

Miyagi agrees to go to the karate school and shocks Daniel by enrolling him in a tournament; Miyagi briefly stepped into the Lawgiver role here. It is very difficult and not advisable to be the Lawgiver, after which you try to become Mentor and Coach. If you're the person that someone perceives as creating their pain, they won't come to you for help. You're the bad guy. If you want to perform these roles, Mentor and Coach, make sure that the Lawgiver is played by someone else.

Miyagi can step out of the Lawgiver role when he sets the tournament up as the new Lawgiver, throwing Daniel into a horrible state of crisis and leaving him with the pressure to get ready. Now Miyagi can offer his help. A comparable example in a business context might involve some people with a report to prepare. You encourage them to present it to higher management and even schedule the meeting. Then, the presentation becomes the Lawgiver, and you're on their side, helping them prepare and learn from the experience afterward.

If a person is actually being ineffective or failing, but believes he or she is doing fine, then there's no potential for change. You must help such people experience their own reality. This is very different from simply pointing out their mistakes. You must create the depth of an emotional experience. With no emotional investment, there can be little learning.

Daniel is now in an even bigger crisis than he'd imagined. He may have won a reprieve from getting beaten up, but now he has the tournament facing him, and he hasn't learned any karate. Now Miyagi begins serious mentoring and coaching, but first Daniel must agree to the rules. Ah, the rules.

Miyagi has invested substantially in Daniel and now demands a commitment from Daniel in return. If Daniel won't give a commitment, Miyagi won't teach him. But Miyagi also knows Daniel has no other options.

Miyagi begins, "Daniel-san, must talk. Walk on road, walk right side, safe. Walk left side, safe. Walk middle, sooner or later, get squished like grape. Here, karate, same thing. Either you karate-do yes, or you karate-do no. You karate do, 'guess so,' squish, just like grape. Understand?"

Daniel replies, "Yeah, I understand." Miyagi continues, "Now ready?" Daniel says, "Yeah, I'm ready." Miyagi says, "Must make sacred pact. I promise teach karate. That's my part. You promise learn. I say, you do, no question; that your part. Deal?"

Daniel gives the only possible answer, "It's a deal." He agrees to the rules, having no idea of the deal he has just made. That's how it always is. Deals aren't made and kept with words but with behavior, so what Miyagi must immediately do is reinforce this deal accordingly. For Daniel, Phase I is over and Phase II begins: training.

Miyagi starts taking Daniel through a rigorous, puzzling, and seemingly irrelevant regimen of activities. This is partially to build the right muscles and reflexes, but is also to get him used to the discipline of training. As Coach, Miyagi helps Daniel learn some vital combat skills without his even realizing it, because the training is disguised. Miyagi has Daniel wash cars, sand decks, paint fences— and Daniel gets real tired of it. He reaches the point where he won't do any more of this. When will he learn karate?

Again, Daniel threatens to break off their relationship; he behaviorally states that he's been pushed as far as he'll go. Miyagi adjusts. He's sharp and knew that Daniel would reach this point, so he's prepared to shift into a new coaching style. Miyagi makes Daniel aware of what his body has learned from all the car waxing, deck sanding, and fence painting. Daniel can't respect this learning until it's made conscious for him, so Miyagi shows him that through each of these tasks, he has actually learned a great deal. With a few punches and kicks at Daniel, Miyagi demonstrates that, without his knowledge, Daniel has learned to block attacks from all directions. Not bad.

Daniel is amazed, and now the expression on his face is that of someone who's ready to learn. Miyagi had taken a deliberate strategy of allowing Daniel to think he was taking advantage of him for a series of home improvements. When Daniel reached the point where he wouldn't put up with it any longer, Miyagi provided meaning.

Miyagi reveals to Daniel that he's always worried about getting in a fight and that he hates fighting. This demonstrates a key characteristic for a real Mentor: telling the truth about his or her personal vulnerabilities. This honesty helps others express their own emotions and vulnerabilities. In American culture, people tend to think if you're scared (and particularly if you're male), there's something wrong with you. Think about what courage is. How could you possibly be courageous if you weren't scared? If you're not scared, where is the need for courage? Maybe what people should do in business organizations is be more willing to tell each other that they don't always know what's going on, that all too often, they don't have all the answers. Americans could all use a lot more honesty.

Daniel finally learns the real purpose of practicing karate: to be able to avoid fighting. This illustrates a subtle and often troublesome obstacle for growth in others and ourselves. Each time people go through a significant growth phase in their lives, a new purpose can be paradoxical from our old point of view. Why would you learn to fight to avoid fighting? One of the largest paradoxes managers work with today is leadership; they're supposed to lead not by leading but by letting go and empowering others. Too often, the result is that managers abdicate any role, believing those empowered should be left alone.

One night, Daniel finds Miyagi kneeling on the floor in his home, drunk. During World War II, while he was in Germany fighting with the American Army, Miyagi's wife died in childbirth at a relocation camp. He has an enormous tragedy lying in his past and has every right to be bitter. Instead of letting it ruin his life, he established a yearly ritual on the anniversary of his wife's death. In the ritual, he reexperiences his grief, lets it wash through him and out of him, and then goes on. He doesn't block the emotion, nor does he let it get in the way of leading a generally happy life.

This episode may not appear to relate directly to the archetype, but it does—and it's a powerful lesson. Celebrations can be rituals of the past that are sad as well as happy. Whether coping with a business disaster or the death of a loved one, people must let themselves experience tragedies emotionally, fully, while not letting them become overwhelming. What Miyagi has done shows great wisdom: he acknowledges his loss once a year, and then he goes on with his life.

In any mentoring relationship, there is a tendency to think the mentoring occurs only in one direction. Sometimes, the roles can be reversed: with Miyagi drunk and weeping on the floor, Daniel becomes the Mentor and cares for Miyagi. Even though he came to Miyagi for nurturing and support, when Daniel finds Miyagi steeped in tragedy, he becomes the Mentor and comforts him.

For his birthday, Miyagi gives Daniel a gift with enormous symbolic power: a karate uniform decorated with a piece of embroidery that his wife had made years ago. Rewards, recognition, gifts—all of these have the greatest value when they carry symbolic power. Tokens of recognition must be real; they must come from the heart and relate directly to the experience that someone has had. People respond strongly to symbols, things that represent shared emotional experiences and the bonds they create.

The movie ends with a blast of joy from Phase III. Daniel wins the tournament, fairly beating all his former tormentors. People in the audience hoist him on their shoulders. He's the Champion, bathing in the spotlight. The Champion is the recognized hero, but Mentors and Coaches get their recognition too—but from the Champion, not the audience. Riding high above everyone's heads, Daniel turns to Miyagi and yells, "We did it!" Miyagi smiles back with satisfaction and pride. See Figure 4.5 for the integration of this learning with the archetype.

BUILDING THE TEAM

The morning after watching *The Karate Kid,* the workshop participants spend several hours talking about the roles, behaviors, and other lessons demonstrated in the movie, many of which are described previously. This is followed by another day of exercises, models, modeling, and coaching. And in the afternoon, the group faces the wall again.

This time, the participants behave very differently. Instead of the haphazard, directionless, and dangerous approach to crossing the electrified wall, their process is now one of prototyping ideas, practicing them in a safe place, and building on first attempts to improve their performance. Only when a technique is satisfactory do they take action at the wall.

Challenging the wall the second time, the workshop groups always break the task into individual challenges and break the group into small subteams to generate ideas for each challenge. They draw, sketch, experiment, and invent ways to make their ideas concrete. Each subteam shares its ideas with the whole group, inviting others to build on their ideas. When they all agree they're ready to try an attempt at the wall, they assign roles: Timekeeper, Wall Watchers, Coaches, and Mentors. Most groups succeed or come much closer to succeeding than they did the first time; not getting all the people over the wall this time doesn't matter. The important thing here is that the people have learned a new process: they've learned the right way to approach a quality issue during Phase II. That's what the workshop is about.

One of the new habits people in the workshop cultivate is building on each other's ideas—learning that one idea is simply an element of the solution, not the solution itself. The participants learn this through the metaphor of building an arch out of bricks to support an apple. As illustrated in Figures 4.6 and 4.7, ideas are bricks or elements and the completed arch is the final solution.[3] As ideas are suggested, more bricks are added to the arch until some final element, working with the others, completes the picture. The arch is complete when the final solution appears.

If you keep this metaphor in your minds when exploring problems and looking for solutions, you can increase the possibility for new ideas, while decreasing idea rejection. People need to convert some of their automatic no's to automatic yes's or at least maybe's. Americans will never be completely free of automatic no's—they're structurally embedded in us. But that's the virtue of prototypes. They're not intended to be complete or perfect, they're intended to invite a "What's wrong with it?" response.

The prototyping process validates and channels our natural automatic no or "bashing" behavior. Once people are free to explore what's wrong with something, they can be tremendously inventive at finding ways to improve it. They'll change their negatives into positives on their own, with no external prompting.

Over time, as a work group evolves from a random collection of people into a high-performance team, they can and will reduce their bashing. They'll find more pleasure, fun, and fulfillment in dis-

---

[3]Edward deBono, author of *Lateral Thinking,* is credited with the bridge of bricks metaphor for creativity.

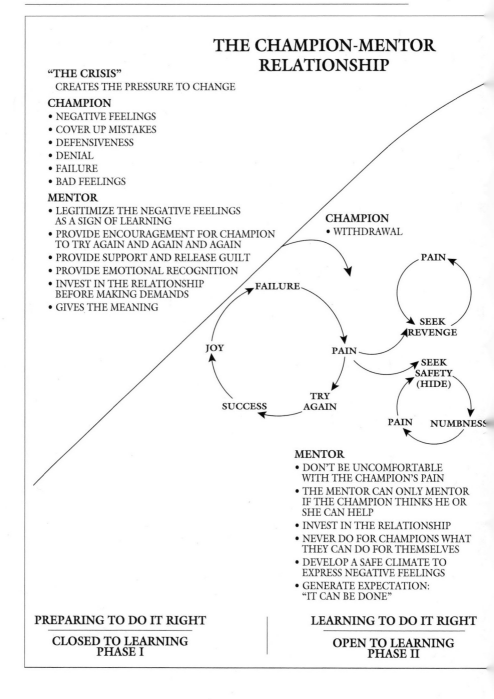

THE CHAMPION-MENTOR RELATIONSHIP

"THE CRISIS"
CREATES THE PRESSURE TO CHANGE

CHAMPION
- NEGATIVE FEELINGS
- COVER UP MISTAKES
- DEFENSIVENESS
- DENIAL
- FAILURE
- BAD FEELINGS

MENTOR
- LEGITIMIZE THE NEGATIVE FEELINGS AS A SIGN OF LEARNING
- PROVIDE ENCOURAGEMENT FOR CHAMPION TO TRY AGAIN AND AGAIN AND AGAIN
- PROVIDE SUPPORT AND RELEASE GUILT
- PROVIDE EMOTIONAL RECOGNITION
- INVEST IN THE RELATIONSHIP BEFORE MAKING DEMANDS
- GIVES THE MEANING

CHAMPION
- WITHDRAWAL

PAIN

FAILURE

SEEK REVENGE

JOY

PAIN

SEEK SAFETY (HIDE)

TRY AGAIN

SUCCESS

PAIN   NUMBNESS

MENTOR
- DON'T BE UNCOMFORTABLE WITH THE CHAMPION'S PAIN
- THE MENTOR CAN ONLY MENTOR IF THE CHAMPION THINKS HE OR SHE CAN HELP
- INVEST IN THE RELATIONSHIP
- NEVER DO FOR CHAMPIONS WHAT THEY CAN DO FOR THEMSELVES
- DEVELOP A SAFE CLIMATE TO EXPRESS NEGATIVE FEELINGS
- GENERATE EXPECTATION: "IT CAN BE DONE"

PREPARING TO DO IT RIGHT

CLOSED TO LEARNING PHASE I

LEARNING TO DO IT RIGHT

OPEN TO LEARNING PHASE II

"WE DID IT"

"THE CHALLENGE"

CHAMPION
- INCREASE IN SELF-CONFIDENCE
- PERCEIVE THE GOAL AS ACHIEVABLE
- REDUCED NEGATIVE FEELINGS

MENTOR
- ACKNOWLEDGE THE DIFFICULTY OF WHAT ALREADY HAS BEEN ACCOMPLISHED
- ACKNOWLEDGE THE ABILITY TO COPE WITH THE REMAINING PROBLEMS
- BE BOTH DEMANDING AND CARING
- WHEN THE CHAMPION HAS PAIN, BE THERE, WITNESS IT, BUT DON'T FIX IT
- TEACH CHAMPION TO TRUST THE QUALITY AND NOT QUANTITY OF WHAT THEY KNOW
- THE MENTOR IS NOT IN THE FIGHT

"THE CELEBRATION"

CHAMPION
- HIGH LEVELS OF POSITIVE ENERGY
- APPROACHING SUCCESS
- HIGH SELF-CONFIDENCE

MENTOR
- RECOGNIZE THE JOURNEY AND ACCOMPLISHMENT
- HELP CHAMPION IDENTIFY LEARNING
- PREPARE THE CHAMPION FOR THE NEXT GOLD MEDAL
- THE CHAMPION WILL TAKE CHARGE

Figure 4.5

DOING IT RIGHT

OPEN TO LEARNING PHASE II

PREPARING FOR THE NEXT CHALLENGE

PHASE III

Figure 4.6

covering ways to make things better. No one will think twice about completing their task quickly and inviting others to help criticize and improve it.

Figure 4.7

PHASE III: CELEBRATION AND PREPARING FOR THE NEXT CHALLENGE

The participants work on their own to design a fitting celebration for their sessions, and this module usually receives the highest evaluation. Without exception, these are wonderfully creative, fun-filled times where every foible of the facilitators, every aspect of the workshop is subject to hilarious, ripping parody. The dinner that follows is often hard to digest because everyone laughs so hard.

After many toasts are proposed and the dinner is over, the facilitators encourage the people to relive their workshop experience. Each person expresses his or her feelings toward the other people in the workshop. This is a time for the participants to express how they have felt toward each other during the time they've been acquainted. They may have just met in the workshop, or they may have known each other for 20 years. As this is happening, a symbolic piece of the wall, their wall, is presented to each person, mounted on a certificate. When all participants have their certificates, they grab colored magic markers and move around the table, signing each other's certificates.

Some readers may have trouble imagining the emotional power of this celebration ceremony. This is the most meaningful experience the people have in the workshop, and one that stays with them forever.

The groups universally describe the workshop as a significant team-building experience; even though this is not the primary goal of the workshop, it seems to be a very consistent fringe benefit. The celebration catapults them to a new level of capability. The greatest lasting effect is the participants' ability to work more effectively as a group and to find and give support to one another. Many other workshops promise this result, but too often can't deliver it. They haven't experienced just a team-building exercise; but they have also learned the three key phases a successful team must go through and what it takes to get through all three phases.

The learning here is the powerful effect that recognition has on people. A celebration, according to the archetype, must bear certain characteristics. It must recognize the work of the individuals and it must tell their story, including all the struggles they've been through. In the workshop celebration, often the most touching stories are exchanged between colleagues who have known each other for years. These stories always include the rough times the celebrants have lived through. Sometimes they mentioned the support, encouragement, and caring someone gave them when things were going from bad to worse, whether on the job or off. The emotional power of this experience moves some men and women to tears—not many, but a few. Such a strong response is to be expected, but is by no means required. The only requirement in such a celebration is sincerity; you must believe and speak from the heart.

— ✧ —

The workshops continue to attest to the validity of the archetype for characterizing Americans' behavior with quality. Because the archetype is sure to be part of American culture for the long term, Americans must find ways to shape their behavior in alignment with its process. Its three phases, and the key roles that must be played for each phase to occur, are the elements of the emotional logic surrounding American quality.

A serious mistake Americans often make is trying to avoid the pain and failure of Phase I. All the books on how to avoid failure contain valuable lessons, but they're lessons that have value only when people are open to learning them. Only by getting through Phase I do Americans become open to learning. All participants in the wall exercise were closed to learning until they saw themselves on tape and could no longer deny their real behavior.

Allowing themselves the full experience of embarrassment opened the door to learning some new behaviors or relearning some old ones. Phase I cannot be avoided. People's first attempts should always be internal prototypes; then they must build on them. Perhaps by the fourth or fifth attempt, they can have something ready for the customer.

The magic of this prescription is that it accomplishes one of Dr. Deming's most important points: driving out fear. The acknowledgment and validation of Phase I reduces fear and makes people willing to take risks. When people systematically include failure in their processes—encouraging it, recognizing it, and celebrating those who try, fail, and try again—they reduce fear and increase risk-taking, and ultimately achieve more major breakthroughs. This is the truth about how Americans work. Although in their hearts they always want to do it right the first time, the fact is that they most often don't and probably never will.

A major problem American managers and workers face is trying to reshape archetypically American behaviors into accepted business school or Japanese molds. Americans must celebrate their uniqueness and build on their strengths, even if one of their "strengths" is not doing it right the first time. This is too often criticized in business but has a vital place in their lives. It fertilizes Americans' explosive, chaotic creativity and their talent for invention and doing the impossible. As James Fallows describes in his book *More Like Us*, the Japanese have a talent for order, while Americans have a talent for disorder. Let's use it.[4]

---

[4]Fallows, *More Like Us*, p. 48.

A problem is a chance for you to do your best.

—Duke Ellington

— ✧ —

Trust each other again. When the trust level gets high enough, people transcend apparent limits, discovering new and awesome abilities for which they were previously unaware.

—David Armistead

— ✧ —

Storms make trees take deeper roots.

—Claude McDonald

5 READY, FIRE, QUALITY

AT THE QUALITY STARTING LINE

Failure, Support, Celebration. The archetype process lives all around us, in massive corporate change programs, medium-sized team projects, and individual efforts. After reading this book, you'll notice it everywhere. You'll see people living through some elements of the archetype and know that other vital elements are missing that might help ensure success. Perhaps you'll reflect on events in your own life and remember living through the pattern in one set of circumstances or another. But best of all, you'll be able to apply the archetype learning to your advantage.

If the archetype lies unconsciously behind American success stories, then consciously repeating its three-phase pattern should lead to more successes. This assertion was certainly borne out in our workshops. For people to effectively transform themselves and their work from a negative to a positive state, all elements must be present: Lawgiver, Crisis and Failure, Mentor, Coach, Support, Champion, Impossible Dream, and Celebration. Once Americans accept the idea that the archetype is the fundamental structure for the human side of quality, they'll find that the number of ways to recreate the pattern is limited only by their imagination.

After divestiture in 1984, AT&T focused for five years on the quality coaching tools in Phase II as the means to continuous quality improvement—without significant corporate-wide improvement results.

However, in 1989 we added a Phase I and Phase III dimension to our quality improvement efforts and have seen significant improvement in that time. Phase I was initiated with the establishment of a Chairman's Quality Award with business units and divisions evaluated annually by the demanding Malcolm Baldrige National Quality Award criteria. This made visible a Quality crisis across the corporation. We were not measuring up to the high quality standards expected in today's marketplace. Our first major, visible nation-wide service outage in January 1990, was an unexpected confirmation of our crisis.

Reluctantly, we accepted the need to improve beyond our internal standards—as the Chairman and the Baldrige evaluation criteria played the role of Lawgiver.

Phase III was initiated with the establishment of a two-day annual AT&T Quality Conference, where the chairman attends both days. Marilyn Zuckerman and Vince Franco were asked to be AT&T Quality Conference co-chairpersons and the annual celebration has become the key AT&T conference of the year. In 1991 over 800 people attended and 45 AT&T locations were on satellite hook-up. The Chairman's Quality Awards for business units and divisions are presented at this conference—along with many Quality Team success stories. The audience and champions, include representation from all parts of AT&T who join together in this annual celebration.

At AT&T the conventional feelings about Phase I, the Chairman's Quality Award, and Phase III, the AT&T Quality Conference, still exist and will not die quickly. But each year more and more people at AT&T are beginning to see the value of Phase I and Phase III—as they experience them.

Phillip M. Scanlan
AT&T Quality Vice President

The literature of modern business offers guidance for Phase II, including the effects of different empowerment strategies and tools and techniques for creating a supportive environment. Much of the material offered under the headings, "Employee Involvement," "Participative Leadership and Management," and "Empowered Employees" provides many of the ingredients needed for success in Phase II. The role of the manager as a Coach is well articulated in many quality books that focus on the analytical and facilitation tools the Coach needs, but the role of Mentor is generally lacking. Phases I and III are rarely discussed. This omission is deadly, for in these phases lie the greatest challenges and many of the vital ingredients.

Elements in Phases I and III create the human energy for continuous improvement and breakthrough results. Unfortunately, today's attitude toward the crisis and failure of Phase I is one of rejection and denial. Common beliefs make us work to avoid Phase I at all costs; it is seen as purely negative, while the conscious wish is to be only positive. On the other hand, Celebration is rejected because it's too indulgent, too pleasurable.

Celebration seems too much like having fun; it's frivolous, uncontrollable, not connected to the bottom line and ultimately not appropriate in the workplace. In business, Celebration is not taken seriously, so it becomes an afterthought, an add-on, a dessert that's probably best left on the plate.

This chapter will look more closely at Phase I and offer stories to illustrate how it works and how you can apply it; the two chapters that follow will provide similar information on Phases II and III. In all three chapters, you'll find definitions of certain key words, common in the conventional language of modern business, contrasted with their new archetype meanings; we hope this will provide a deeper understanding of their subtlety.[1] You'll also find a rich source of ideas to help you jump into action.

CLOSED TO LEARNING

Probably the greatest challenge related to Phase I is simply accepting the fact that you can't avoid it. Americans must learn that Phase I is the first critical step, one that holds equal opportunity for ultimate success or failure, depending on how they manage it. This phase is integral with the archetype's learning process, just as much for adults who repeat it as for children who experience it for the first time. See Figure 5.1 for the contrasts between the conventional and the archetype views of failure.

---

[1]The language in the figures describing the difference between the conventional meaning of words and the archetype definition was developed jointly by the authors and the American Quality Foundation.

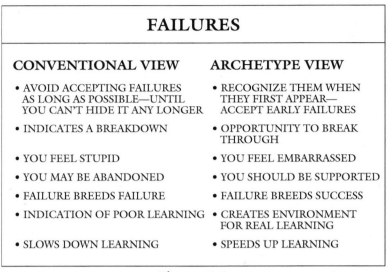

FAILURES

CONVENTIONAL VIEW	ARCHETYPE VIEW
• AVOID ACCEPTING FAILURES AS LONG AS POSSIBLE—UNTIL YOU CAN'T HIDE IT ANY LONGER	• RECOGNIZE THEM WHEN THEY FIRST APPEAR— ACCEPT EARLY FAILURES
• INDICATES A BREAKDOWN	• OPPORTUNITY TO BREAK THROUGH
• YOU FEEL STUPID	• YOU FEEL EMBARRASSED
• YOU MAY BE ABANDONED	• YOU SHOULD BE SUPPORTED
• FAILURE BREEDS FAILURE	• FAILURE BREEDS SUCCESS
• INDICATION OF POOR LEARNING	• CREATES ENVIRONMENT FOR REAL LEARNING
• SLOWS DOWN LEARNING	• SPEEDS UP LEARNING

Figure 5.1

While in Phase I, people are virtually closed to learning. They tend to engage in certain counterproductive behaviors, like hiding, seeking revenge, or shooting the messenger. Given conventional management practices, such behaviors are extremely difficult to avoid. Regardless of the size of a project or the number of people involved, Americans tend to set a fixed course and follow it until an unexpected problem occurs. If the problem can be concealed, the first reaction is: "No one else has to know, we'll take care of it." The longer people follow the original plan before trouble appears, the more emotionally invested they become in the decisions made along the way, and the less willing they are to see other possibilities. When they receive criticism, they defend their decisions with astonishing inventiveness, creating alibis and finding other targets to blame.

Everyone has his or her share of Phase I battle scars. Consider those times when you were supremely confident in your understanding of a new project. Some time later, it all blew up, and you learned, to your embarrassment, how little you had really understood it. Or maybe it didn't totally destruct; maybe only a few, easily concealed defects appeared. This situation is more common. It's initially less embarrassing but personally more risky. You can keep quiet about the problems, send the product out, and hope for the best or choose to reveal the truth, risking not only embarrassment but also personal rejection.

In the latter case, you fight the good fight and preserve quality; but you will also probably create delays, fall under management scrutiny, and let everyone know you're not perfect. Today, one of management's most important tasks is creating a working environment in which people feel safe taking such personal and professional risks.

SHOOTING THE MESSENGER

Predictable Phase I behavior appears everywhere. When problems occur, people tend to hide them or blame others. Instead of dealing with the real causes on the human side, those charged with tracking down the culprit generally attack the symptoms: faulty software, antiquated machinery, or budget cuts. In America, even the most incisive root-cause analyses of major problems typically deal only with technical issues. Americans are much more comfortable looking for something like flaws in the structure of a rocket engine seal than they are examining the human process of how an organization handles bad news, even though such a process is often the real root cause. This behavior occurs at some point during the life of every project, and the earlier it shows up the better the chance for success. When problems occur toward the end of a project, they're often large, complex, and unwieldy, the accumulation of many smaller problems or errors that were successfully hidden and left uncorrected.

As Dr. Deming has pointed out, people commonly hide problems for one reason: they're afraid. In *Driving Fear Out of the Workplace*, Kathleen Ryan and Daniel Oestreich define fear in the workplace as "feeling threatened by possible repercussions as a result of speaking up about work-related concerns."[2]

They define an undiscussable as "a problem or issue that someone hesitates to talk about with those who are essential to its resolution . . . [and which] represents a potential barrier to doing quality work or building an effective work environment." In their extensive

---

[2]Ryan and Oestreich, *Driving Fear Out*, p. 21–26.

research on fear in the workplace, these authors identified management practice as the leading cause of fear and the primary undiscussable, far outweighing any other item by at least four to one. When exploring the consequences of taking action despite these fears, they found that the single thing people fear above all else is "loss of credibility or reputation . . . . This includes being seen as a troublemaker, boat rocker, agitator or not a team player, or being given other labels that mark the individual as a problem to the organization . . . [and] includes fear of losing influence or of being seen as not possessing good judgment or acting in an unprofessional way."

One thing that can help organizations combat this situation is a spirit of honesty and openness at the highest levels. Everywhere, companies are being destroyed from within by the shoot-the-messenger syndrome, a source of crippling and often well-founded fear in the workplace.

In part, this attitude grows from an executive's habit of residing in an ivory tower and avoiding being associated directly with anything that hints of the slightest imperfection. It's not surprising that lower-level employees are reluctant to expose problems or failures when their organization's leadership always conveys an attitude of being above imperfection. Studies have revealed that when senior managers and executives express their own problems, fears, doubts, and past failures, they become more vulnerable and credible as human beings; this also makes them more approachable when bad news is an issue.

Rather than issuing "From the President's Desk" statements and chatting with employees about nothing but success, managers are advised to sprinkle their success stories with talk about their past failures, their troubling doubts, and the bad decisions they've made during their careers. In doing so, they become less feared and more admired. When they acknowledge their human journey, and model the human side of quality, they grow closer to their people and help create a more honest environment throughout their organization. This is the first step toward building a true learning organization, one in which it's okay to make mistakes and easy to spot problems.

While the role of the boss, traditionally the Lawgiver, is beginning to change to a combined role of Mentor and Coach, the Lawgiver role as defined by the archetype must still be played. See Figures 5.2 and 5.3 for the contrast between the conventional and the archetype views of the Lawgiver and Crisis. The boss may still have to be sure that the voice of the new Lawgiver, the customer, or competition is the "right" Lawgiver and is effectively heard and acted on by the organization.

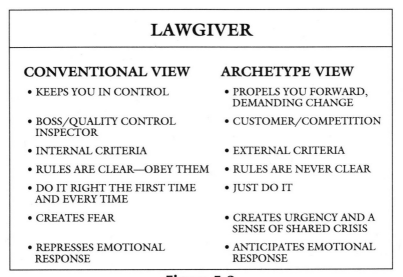

LAWGIVER

CONVENTIONAL VIEW	ARCHETYPE VIEW
• KEEPS YOU IN CONTROL	• PROPELS YOU FORWARD, DEMANDING CHANGE
• BOSS/QUALITY CONTROL INSPECTOR	• CUSTOMER/COMPETITION
• INTERNAL CRITERIA	• EXTERNAL CRITERIA
• RULES ARE CLEAR—OBEY THEM	• RULES ARE NEVER CLEAR
• DO IT RIGHT THE FIRST TIME AND EVERY TIME	• JUST DO IT
• CREATES FEAR	• CREATES URGENCY AND A SENSE OF SHARED CRISIS
• REPRESSES EMOTIONAL RESPONSE	• ANTICIPATES EMOTIONAL RESPONSE

Figure 5.2

PROTECTIVE INSTINCTS

From published reports, it appears that in the series of events leading to the *Challenger* disaster, a widespread preference for fiction over truth contributed far more to the growing snowball of problems than did the solid rocket boosters' defective O-rings. The first concern of the people in charge was the schedule, and they refused to listen to any information that might jeopardize it. Although it was possible to fix the seals, gaining agreement to fix them at the cost of missing their launch date was not. Only after human deaths, the ultimate failure, was the project stopped to resolve the O-ring problem; it might have been corrected if the human processes hadn't grown so unresponsive.

In environments where failure is outlawed, big, visible failures bring people's remarkable alibi creativity into action. After the

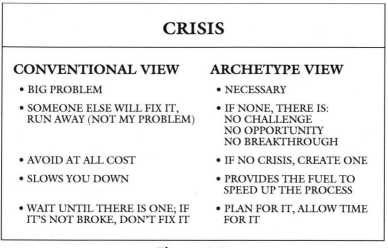

Figure 5.3

Challenger disaster, the investigating committee spent much energy dealing with smoke screens, meandering down long trails of technical explanation. When the investigation was nearly closed, committee member and physicist Richard Feynman stepped forward with a demonstration of first class boat-rocking. He blew away the technological mumbo-jumbo by placing a sample of O-ring material in a cup of ice water and pointing out, a short time later, that the material had become quite brittle. This simple demonstration made it concrete and inescapable that the root cause of the disaster was not technological: how could anyone involved in the design of the shuttle not know this fact about the O-ring material? Feynman forced the committee to deal with the painful reality of NASA's concern with meeting its schedule, and for this he was despised by many of his fellow members. The explosion had already forced NASA and the managers of Morton Thiokol, the makers of the solid rocket boosters, to relinquish any hope of keeping the shuttle program on schedule. Now Feynman had taken away their alibi: it was no longer possible to assign blame to a technical bug as a means of protecting the reputation of the people involved.[3]

[3]"Last Journey of a Genius," January 1989.

The Bay of Pigs incident in 1962 grew from similar problems. The plan called for a group of Cuban nationals, trained by the CIA, to invade Cuba, trigger a grass-roots uprising, and ultimately over-throw Fidel Castro's government. This was not a back-office or lower-echelon plan—it was developed and discussed by the top advisors in the Kennedy administration.

The invasion failed. The CIA-trained forces were killed or imprisoned, creating an embarrassing international incident for the United States. President Kennedy couldn't understand how he had made such a horrible blunder. The answer is that his advisors had let him. They knew the invading forces would be outnumbered 140 to 1 and that the local population would almost certainly not rise up to support them; but they were afraid to tell him. They sub-stituted their belief in possible success for the truth about probable failure. They didn't dare spoil the plan or jeopardize their status as members of an elite group.[4]

People in situations like this create alibis to protect themselves from pain when the truth disturbs or threatens their sense of emo-tional security. People need to feel safe and want to be liked and well regarded. They are quite ready to hide a piece of truth to protect these interests. The tendency is even stronger when people act as part of a group: a family, design team, management committee, or, as in the Bay of Pigs case, the National Security Council. The more powerful the group, the more people want to remain a part of it and the greater their need to conceal information that might jeopardize their membership status. Like parents trying to protect their children from an unpleasant challenge or painful emotional blow, all people have a remarkable, almost knee-jerk desire to protect themselves and their fellow group members. When Lew received word that the Copper Shop might be shut down if its bottom line didn't improve, he initially kept this knowledge to himself, even though this choice meant he had to bear a terrible emotional burden. At least for a while, it was preferable to being the bearer of bad news.

In time, alibis become part of the regular routine of work, like an in-joke that everyone smiles at but never speaks out loud. In this way, the habit of deception can be perpetuated for decades and

---

[4]Bill Moyers, "The Truth About Lies," November 1989.

spread through organizations of thousands. By inadvertently encouraging and perpetuating this attitude of protective deception, people avoid the short-term anxiety of confronting the painful truth but never strengthen themselves or their organizations. Instead, they allow the painful truth, whatever it might be, to remain hidden. The problems begin to snowball, becoming more difficult to hide, and people spend more time and energy trying to hide them. In time, people's willingness to deal with problems honestly and effectively becomes almost nonexistent.

FIRST ATTEMPT

As with words like control and perfection, many people react against the words that best describe Phase I: Crisis and Failure. To make Phase I more acceptable, it's easy enough to give it a different name. Instead of Crisis or Failure, you might refer to it as the Prototype Phase, the First Attempt, or Take 1. Tom Peters noted that many people in his seminars could not even say the word failure. The closest they came was "the hated 'F' word," or "outcomes of the other variety." He makes some very applicable recommendations in *Thriving on Chaos:*

> *The goal is to be more than tolerant of slip-ups . . . actively encourage failure. Talk it up. Laugh about it. Go around the table at a project group meeting or morning staff meeting. Start with your own most interesting foul-up. Then have everyone follow suit. What mistakes did you make this week? What were the most interesting ones? . . . Make it a habit to send thank-you notes to people who make innovative, fast failures; send such a note around the office.*[5]

Rather than invest financially and emotionally in trying to avoid this phase, literally trying to do it right the very first time, people are far better served when they accept initial failure as an advantage, an essential and valuable period of experimentation and learning: the first attempt.

---

[5]Peters, *Thriving*, p. 261.

As mentioned in Chapter 4, Americans are action-oriented; planning plays second fiddle to doing. The logical approach of plan first and do second is overridden by the unconscious pattern of the archetype. The antidote for this inescapable pattern, and its challenge, lies in finding ways to use it to achieve the desired results.

Accepting that first attempts rarely work properly stimulates greater risk-taking, creativity, and learning. Combining Peters' fast failure ideas with the archetype's first attempt can change the dynamic; some might say it could shift a paradigm. This shift creates the opportunity for people to be inventive and experimental, because their fear of failure has been substantially reduced.

The first attempt does just that. It's an opportunity to approach any project in a way that simply lets people do it, and as quickly as possible. Whether you're creating a product or a service, for an internal or external customer, try to simulate the full scope of the project and have people execute a fast failure. The purpose of the first attempt is to open people to learning, to prepare them for Phase II. The first attempt demonstrates what they are able to do, based on their initial understanding and capabilities. The faster they execute it, the less ego-invested they are in its outcome.

After completing the first attempt, people are ready for feedback. While feedback from others is valuable, the best feedback is each person's own observation of the results. Not only do they learn what needs to be done to improve it, to bring it to the desired state, they also learn what they don't know. When they see and experience their own first attempts, they make their work, and how they feel about it, concrete.

Giving and receiving feedback about work in process is a critical component of quality improvement, but typically the giver and receiver in this process feel some degree of uneasiness. When those receiving feedback have invested substantial time, creativity, or personal risk, they are more vulnerable to another's judgment. It can cause them tremendous pain and even lead to hostile counterattacks. The greater the receivers' defensiveness, the less likely it is that the givers will provide all the feedback they have to offer. If the givers fear any sort of reprisal for being honest, they'll be reluctant to provide any negative feedback at all, however constructive it might be.

First attempts with fast failures can help encourage an environment in which beneficial feedback occurs more freely. After the producers (of fast failures) have experienced their own role as critic and have seen the flaws in their work, they are more open to criticism and suggestions from others. In an environment that encourages failures (and learning from them) the producer is more likely to make the critics feel comfortable, to encourage them to give all the feedback they have, no matter how negative. This attitude represents a complete, 180-degree reversal, from hiding failures to shining a spotlight on them. In the environment geared toward fast failures, people just might happily rip open projects to find every flaw. In some cases they will even force things to fail, so that their quality in the second attempt will be far greater.

People should consider any and all ways to help each other when working on fast failures. They should try to find ways to experience their own results as if they were customers, and in so doing become better critics. A video camera and recorder are wonderfully effective tools for self-criticism. Watching the video of such first attempts is a powerful way to open people to the need for learning, seeking out coaches, and perhaps attending skill-based training sessions. The advantage here is that such activities are self-directed rather than being suggested or imposed by a manager. For real growth and development, this can make all the difference in the world.

MEASURE TWICE . . .

"Begin with the end in mind."[6] This bit of wisdom is based on the principle that all things are created at least twice. The first creation is in the mind of the creator; the second makes a thing physical. And there should be a third creation, and a fourth. Long before a product or service enters the marketplace, customers should be involved in these early attempts to provide first-hand information and criticism on how the thing you're trying to produce relates to what they really need.

Consider the construction of a house. Architects create houses in every detail before the first nail is hammered into place. Even

---

[6]Covey, *Seven Habits,* p. 99.

before it's put on paper, they must have a clear sense of the kind of house it should be. Engineers, strategic planners, designers, poets, artists, and musicians—these and many other professionals work with ideas, with their minds, until they get a clear image of what they should create. And they often apply the carpenter's rule: "measure twice, cut once." In other words, create a prototype internally before you release an alleged "finished" product externally. It's another perspective on fast failure.

Before constructing a building, architects build models, physically or simulated with a computer. When chemical companies develop new products, they go through a prototype phase tailored for their industry before beginning mass production: they construct a pilot plant. This is an intermediate stage between the research lab and the factory, which brings an experimental process one step closer to commercial viability. It produces the new material on a larger scale than the lab but doesn't require the investment, risk, and public exposure of full production. A pilot plant lets the engineers identify manufacturing, contamination, or control problems that would never have shown up during the research phase but that would have spelled disaster for mass production.

The question is not whether to try to avoid initial failure—you can't. Rather, it is whether to let the failure occur internally, as an acknowledged pilot project or prototype, or externally, in the hands of your customer who becomes an unsuspecting victim. The former gives you control, protection, and the ability to trap and correct problems very early in the product's life cycle. The latter forces the customer to suffer through lower-quality products or services and forces people into a frenzy of patching holes and fixing problems that never should have reached the customer in the first place. This situation can damage an organization's reputation, sometimes irreparably.

The people developing new products and services should work as closely with their customers as possible, involving them early in the process, listening to their opinions, and letting them play an active role in creating what they'll eventually use. Rather than treat your customers like guinea pigs, give them all lab coats and make them part of your earliest efforts.

Some organizations pursue an effective compromise between internal prototyping and full-blown external marketing; they initiate small, fast, inexpensive trials for new offerings. This approach enables them to gather information about customer response very quickly, without the risks connected with a national product launch. The key to success here is not worrying about the accuracy or effectiveness of the first effort. What you're after is the accumulation of knowledge from many customers, reflecting their reactions to a number of different marketing approaches or product designs, and you want this knowledge as fast as you can get it. Begin a rapid-fire series of marketing efforts. As information comes back from the customers, the marketing strategy and the product or service itself can be adjusted accordingly. In the words of C. K. Prahalad and Gary Hamel, "Little is learned in the laboratory or in product-development committee meetings. True learning begins only when a product—imperfect as it might be—is launched."[7]

No matter who you follow as a business and management guru, all the findings about initial failure are basically the same, and our cultural research bears them out. Whether you see it as a chemical engineer's pilot plant, Tom Peters' fast failure, or an application of Prahalad and Hamel's low-cost, expeditionary marketing, you're seeing a reflection of Phase I of the American quality archetype; any new project begins with and benefits from failure. This is the time to learn, gain energy, and refine the prototype into something of unparalleled quality. The wisest course for today's manager is simply choosing the best method for initial failure in his or her particular organization. Remember, the faster the failure, the faster you learn how to improve.

LEW TRIES A PROTOTYPE

Doug is one of the managers in my organization. We have many goals in common, one of which is to support the growth of people. As often as possible, we try to look beyond the more conventional managerial role of shaping their professional growth and try to help

---

[7]Hamel and Prahalad, "Corporate Imagination," p. 81–91.

them achieve success on a personal level. Doug and I believe that a person who feels cared for brings energy, enthusiasm, and creativity to the job. The following story explores these efforts in connection with a fast failure. It involves two people: Sherl, a management associate, and Madeleine, a senior engineer.

At a point when Doug and I were more swamped than usual, we received a call from a member of one of our sales teams: a customer had invited us to participate in their annual "Supplier of the Year" award. We were delighted with the news but also faintly panic-stricken at the thought of the effort required to participate. To be considered for the award, we had to submit a complex, detailed application; the process was a lot like applying for the Malcolm Baldrige National Quality Award but on a smaller scale.

It looked like four to five weeks of work to gather data, assemble it in the requested format, and perform the necessary reviews. Still, we saw the value in participating, and despite our workload, decided to give it our best shot.

Normally, Madeleine, Doug, and I would handle a project like this; but Doug and I would be on the road for several weeks right in the middle of the application schedule, and Madeleine would be on vacation part of the time. We decided to bring in someone else from our department to work with Madeleine, a person who had the skills to manage the project but would also see it as a challenge and a growth opportunity; Sherl was selected. She had grown considerably during the past several years, but we'd never asked her to participate in anything like this. We felt that, with our support, she would overcome any problems that might arise.

After discussing the project with Sherl and Madeleine, I got the feeling that Sherl was a bit unnerved by its magnitude, even though she never expressed it openly. I had decided to use a fast-failure approach and thought it best to tell them why. Sherl and Madeleine had only been exposed to general archetype concepts, so I filled them in with some additional background that related more specifically to this project; I stressed the points that quality is an emotional experience and always begins with negative emotions. Next, I explained the role of failure, noting how it creates the energy to move forward, and finally discussed how I planned to use this learning to approach the award application.

We would do a fast failure, what I called "Attempt 1." In a day and a half, we would do our best to put together some sort of award application and then review and critique it. We would all think of it as absolutely nothing more than a fast failure. We would follow this with "Attempt 2," a second day-and-a-half prototype. Several such failures would get us up and running quickly. We would all learn from our mistakes and from the ideas that didn't work.

I explained that they should try to make the fast failures as complete as possible and that they'd be free to do whatever they thought made sense. The looks on their faces said it all: panic.

I put my cards on the table and told them frankly that the beginning of this project would probably produce some unpleasant feelings. Most likely, they would feel thrown into a crisis: confused, upset, maybe even angry. But I made it very clear that these feelings would create the energy needed to move into action. Because of the short time frame permitted for a fast failure, they wouldn't have the normal opportunity to become emotionally attached to their work and wouldn't become too defensive when we tore it apart in our first review session. Instead, they would realize it was only the first attempt we were dissecting, not them or their ability. This approach would quickly identify what we knew and didn't know and would enable us to bring together the necessary resources to close the gaps. Sherl and Madeleine nodded approvingly but were clearly concerned and somewhat anxious about this strange, managerial voodoo.

I asked Sherl to keep some notes about her feelings during the project, so we could discuss them at its conclusion, and hopefully share her experience with the rest of our team. I saw this as an opportunity for her to relive the entire journey, the highs and the lows, and by so doing reaffirm her worth, to herself and to us. If done correctly, in the spirit of an archetypal Celebration, I felt it might also inspire the other team members. Sherl agreed.

Attempts 1 and 2 were executed very quickly. Within days, we had a much better understanding of how the application would be handled. With a great deal of hard work, Sherl and Madeleine com-pleted the final application in less than three weeks. Much later, when the "Supplier of the Year" customer reported that our work was outstanding, we all felt the fast-failure approach had worked better than expected.

After the application was submitted, Sherl and I met to discuss the project. She reported that everything had happened just as I said it would.

When she returned to her desk after our first meeting, she settled into an anxious state of panic, seriously doubting her ability. She spent the little time remaining that day discussing her dilemma with a coworker and then mulled it over again that night with her husband. Both offered support and assured her that she could do it. She knew she didn't have much time to think about personal concerns, because of the day-and-a-half deadline. Suddenly, she became caught up in the activity. She knew I didn't expect the first attempt to be right anyway, so she decided to take some risks, break new ground, and use her creativity.

By the end of Attempt 1, her confidence level was higher. Her performance and her attitude were much better than she'd hoped—she was making a contribution and helping us reach an important goal. Having critiqued her first attempt herself, Sherl was comfortable receiving criticism during the first project review, in which we all discussed the positives and negatives in a constructive way. Because we were finding them early in the project, we addressed the problems in her approach quickly and easily. Some of the problems might never have surfaced if we had proceeded traditionally, working for a week or two and trying to do it right the first time. In that scenario, and with that time frame, she would have become emotionally attached to her initial ideas and would have had a hard time changing direction and seeing her work torn apart. This was how she'd always felt in the past when someone found fault with her work.

Sherl also felt that management (Doug and I) had shown a lot of interest and concern for her well-being; she appreciated our frequent reassurance. She was glad that we'd asked her to participate in this project and that we openly talked about her work once it had begun. This convinced her beyond any doubt that we saw her as a valuable member of the team.

When Doug and I complimented her on the great job she and Madeleine had done, she beamed like an Oscar winner. Her reward, she said, came from learning about what it takes to complete a project of this magnitude. She also gained a greater perspective on the business and felt the training she'd received over the past few years had prepared her to take on this assignment.

These are her comments:

> *I liked the fast-failure approach because it did give a sense of urgency to the project. Instead of going back to your desk and pondering over it a few days you jump into action because it has to be done. I think I do thrive on chaos in a way.*
>
> *Well, I think after Attempt 1 I felt better, a lot better. Because I felt like my work was good. Although I didn't have time to complete the task, what I did I thought was pretty good. And I wanted to get the feedback, getting it made me feel good. Attempt 2 was a little different. Attempt 2 made me more nervous than Attempt 1 did. I think maybe because there was more feedback and section two of the application got changed completely. Also, I knew Madeleine was leaving and it was getting closer to the actual time when the application was due.*
>
> *At the time I didn't feel as though we had made enough progress. I was more nervous. I was kind of up, down, and up, and I did get irritated with some comments that were made late in the game, probably because I was emotionally attached to the work. But I feel good about the project and that you and Doug gave me the opportunity. And I feel good that I did it. And as far as recognition, just saying that I did a good job makes me feel good. That kind of thing helps me a lot.*

PHASE I . . .

Accepted as a critical component for success in America, Phase I holds the key to unlocking people's energy and putting them in the lead. Denied or suppressed, it will continue to haunt them. At the very least, denial of Phase I will bring morale problems, low productivity, and dissatisfied customers. At the worst, it will bring more oil spills and air crashes.

Put Phase I to work and use it to your advantage. Inaugurate fast-failure approaches that prevent people from becoming too well defended against feedback. Let people draw emotional energy and enthusiasm from their initial failures. Do quick-and-dirty internal prototypes to force problems to the surface and invite customers to participate in the second, third, or fourth attempts. Approach Phase I in any way that fits your organization—but whatever you do, don't pretend it can be avoided. This is one instance in which you'll never beat the system.

IDEAS FOR ACTION

PHASE I – CRISIS AND FAILURE

Crises create energy for movement. Don't suppress them, focus them. Don't "protect" the people by shielding them from a crisis, share it in a dramatic way that touches people emotionally.

– ✧ –

Think of your prototype as "Take 1" or "First Attempt."

– ✧ –

Eliminate the "F" word: fear.

– ✧ –

Learn to love the other "F" words: fast failures and first prototypes.

– ✧ –

We understand when we do. Simulate the result fast to make it tangible. This helps you learn what you don't know so that you can improve it beyond what's expected.

– ✧ –

Ensure that Lawgivers, Mentors, and Coaches are available.

– ✧ –

Failures breed success. Don't hide them.

– ✧ –

In your workplace and home, make sure people are safe to make mistakes. Recognize people who identify problems before solutions are known. Remember Hub's Yellow Flyers in Chapter Three.

Look for the opportunity in every problem. When problems cease, so do opportunities.

– ✧ –

Don't suppress emotions. Powerful learning grows from powerful emotions. Nothing worthwhile is ever learned in ten easy lessons.

– ✧ –

Seek out crises. They provide the human energy for breakthrough results.

– ✧ –

Never dwell on what you have lost. If you do, you may become discouraged and defeated.

– ✧ –

Admit your weakness and talk about your mistakes, when appropriate. You will gain respect and understanding.

– ✧ –

Recognize failure as a normal part of life. Don't try to escape from it. Instead, accept it as a form of corrective feedback.

– ✧ –

Once faced with a failure, never look at what you have lost. Ask yourself what you have left and what new possibilities are open to you as a result of the failure.

– ✧ –

When faced with adversity, tell yourself "I can do it, I will do it, my dream is possible." Use your imagination and creativity to get yourself unstuck.

Create your equivalent of Hub's Yellow Flyers to recognize people who identify problems, mistakes, or failures.

– ✧ –

When the crises hit, let everyone know you're right on schedule.

– ✧ –

Hold full-team rather than one-on-one meetings to share and deepen the emotional response to the crisis.

– ✧ –

Create prototypes often, first on the entire project and then on each successive phase. Each prototype should simulate the end result with as much detail as your knowledge and capability will allow and remember; do them fast.

– ✧ –

Use video cameras often to provide individuals and teams immediate, concrete feedback on their first attempts.

*A great coach doesn't try to change a great player.
Instead, the coach discovers what is unique, what is
great, about people and honors it, is happy for it, uses it.*
—John Bunn

— ✧ —

*The years forever fashion new dreams when old ones go.
God pity a one dream man.*
—Robert Goddard

— ✧ —

The future belongs to those who believe in their dreams.
—Eleanor Roosevelt

6 GATHERING SUPPORT

OPEN TO LEARNING

There's much excellent business literature available offering examples and how-to advice on Phase II: Support. Instead of repeating what many of these sources have already said quite well, this chapter will focus only on the elements of Phase II that are most important: caring, Mentors, Coaches, and Impossible Dreams. We'll begin with a discussion of that most elusive of archetype concepts: caring.

Today, there's a subtle but powerful shift occurring all around us. People are changing at all organizational levels, from top executives to front line personnel, in every type of organization: corner grocery, government agency, public school system, and Fortune 500 corporation. Some groups are changing faster than others, but all are shifting their stance, expanding or shrinking in response to the economic and social pressures of the late 20th century. Externally, people are pressured by competitive threats; they wonder how much more market share they'll lose or how they'll be affected by new government regulations.

Internally, people are pressured by emotional struggles to protect themselves and loved ones from pain and unhappiness and by moral struggles that result from trying to do what's right. They feel the need to be simultaneously true to themselves, their families, and their organizations, a three-way balance that, for many people, is becoming virtually impossible. In all walks of life, people feel one or more of these forces directly and experience the others indirectly.

Today, Americans are steeped in crisis, feeling the impact of well-publicized trends in economics, politics, the environment, and a wide range of social issues. They feel the presence of crisis in their families and workplaces, hear about it relentlessly through the media, and talk about it with friends. The stress of living in crisis-ridden organizations causes them repeatedly to question the purpose of life, work, or family and to wonder why the old tried-and-true practices of daily life aren't working; they yearn to be enriched by the experiences of life rather than depleted by them.

Everyone rides the same roller coaster, reaching highs and lows at different times for different reasons. All people, at one time or another, come close to giving up. Likewise, all people have experienced the excitement and jubilation of those rare magical days when everything and everyone comes together perfectly, yielding undreamed of results. If you reflect on the course of your life, you'll probably realize that your most important and meaningful learnings came not from the brightest hours but from the darkest. And you may also realize that at most of those moments, when your energy reserves were drained and you were close to giving up, someone was there for you. It was often someone you weren't even particularly connected with, just someone who reached out and cared. This is the essence of Phase II.

Caring has two faces: the empathic Mentor and the demanding Coach. In recognition of the importance of Mentors, many corporations have established mentor programs. Typically, more senior associates are assigned as Mentors for the relatively junior associates, who are often people seen as having executive potential. Such programs and practices do not follow the spirit of the archetype; mentoring is not simply a matter of helping someone advance his or her career. These programs are neither appropriate nor effective for providing helpful mentoring according to its archetype definition. See Figure 6.1 for the contrasts between the conventional and the archetype views of Mentors.

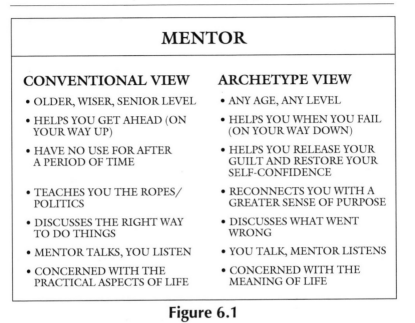

Figure 6.1

Mentors show that they care in many ways; sometimes, it's a very small gesture. Imagine a sales presentation in which you've chosen to take some risks, hoping to knock the customers out of their chairs. Instead, they eat you alive, alternately ripping your data to shreds and yawning at your meticulously detailed graphics. You feel like running and hiding, when someone across the conference table catches your eye and in one look conveys support and respect, even though the room is filled with attackers. You realize this person cares about you, and it makes all the difference in the world. Sometimes caring needs to be demonstrated in a big way. When an entire project is in jeopardy, caring takes focused mentoring of many people, spread over a long period. It may be necessary for the Mentor to make demands that bring out the best in people.

Perhaps you can think of a time when you would have benefited from having a true archetypal Mentor; maybe you need one now. You may have someone on your staff who needs the support of a good Mentor, or you may have entire teams that have recently experienced a setback or defeat and are stuck in counterproductive Phase I behaviors. These are individuals and groups who need you to care and to care enough to demand their best. They would benefit from a Mentor's abilities to listen and to give. If you haven't yet invested emotionally in these people, now is the time to give and give generously. Giving does not mean giving advice. It's giving your concern and attention that will help people. It's caring and listening with your heart.

FROM THE HEART

A charming and insightful book has recently joined the ranks of business literature: *Managing from the Heart* by Hyler Bracey, Jack Rosenblum, Aubrey Sanford, and Roy Trueblood.[1] It offers a concise, understandable description of mentoring behavior, beautifully framing one side of a manager's job. Today's managers need to be Coaches as well as Mentors.

The book is structured as a fictional story about Harry, the tough-minded, strong-willed general manager of a large petrochemical plant. One day, Harry suffers a fatal heart attack, but the powers that be in the afterlife decide to give him a second chance. Like George Bailey in *It's a Wonderful Life*, he's taken under an angel's wing and given some valuable advice.

The angel, cryptically identified as the Woman, tells Harry in no uncertain terms that in return for his second chance at life, he must get his managerial act together (Harry is a steadfast believer in management by intimidation). The Woman becomes his Mentor and teaches him five key rules, each one expressed as an employee's plea for caring and support:

H ear and understand me.
E ven if you disagree, please don't make me wrong.
A cknowledge the greatness within me.
R emember to look for my loving intentions.
T ell me the truth with compassion.

We found these points to be excellent guidelines for mentoring and offer the following comments to expand on the meaning of each.

---

[1] Bracey, Rosenblum, Sanford, and Trueblood, *Managing from the Heart*, p. 18.

Hear and understand me.

When people try to communicate with you, it's important that they feel you're really listening to them; this is one of the cornerstones of trust. Even when you think you know what people have to say, you should still let them express themselves first. They may very well surprise you.

Stephen Covey advises, "Seek first to understand, then to be understood."[2] When you listen to someone, your mind is usually working overtime, actively forming a response to whatever is being said. Covey suggests a better habit is to focus first on fully understanding what is being said. When, as a listener, you seek to understand fully before offering a response, you create trust in the conversation, and the speaker is more likely to feel appreciated and respected. Having been heard and understood, the speaker is more likely to be open to hearing and understanding a response that opposes their initial comments.

As a manager, you may have lots of great ideas; this is fine. But if you have a habit of relating your brainstorms as if you've found the ultimate solution to a problem, you may be cutting off your greatest source of information: other people. You'd be amazed at the good ideas people can have. But if they think you're always completely satisfied with your own opinions, they're not very likely to offer one of their own. Always maintain an attitude that lets people know you're open to their ideas.

During discussions, respond to people with occasional summaries to register your understanding of what they've just said. This gives them a chance to clear up any misunderstandings and also demonstrates that you're actually paying attention. It's important that you do not repeat their words exactly; a robot could do that. Paraphrase their words using your own, enough to let people know they've gotten the message across. This way they'll know you're listening and honestly trying to understand. When it's your turn to speak, they'll be ready and willing to hear what you have to say.

---

[2]Covey, *Seven Habits*, p. 189–190.

Even if you disagree, please don't make me wrong.

People make their own decisions, which don't always coincide with yours; a person charged with designing a new order-entry form may have a very different idea of what it should look like than you do. You may want the person to scrap his or her approach and take direction from you, but that doesn't mean the person didn't try.

When telling people they've made mistakes, or that you want them to do something differently, never question their basic worth as human beings. This may sound like something you wouldn't do in a million years, but if you've ever insulted someone, that's exactly what you've done. If you brand someone with a negative label like, you're stupid, you're naive, or you blew it, then you question that person's basic worth. The person will feel it and resent it. At the very least, he or she will feel hurt and depressed; at the worst, angry and vindictive.

This doesn't mean you can't discuss your concerns and disagreements. Just make sure your discussion is a two-way communication, not a one-way attack. Instead, talk in terms of your reaction to what a person has done: I think there might be another approach; I would have done it this way; or I have a different idea. Statements about yourself don't create as many bad feelings; it becomes an issue of your reaction rather than their failure, and you still get your point across.

If the situation feels threatening to you, try to avoid projecting this feeling as a threat to the other person. If this happens anyway, admit it, apologize, and ask to try again. This time, concentrate on the first principle.

Acknowledge the greatness within me.

We all have the potential to grow, to expand our achievements and our understanding of life, but more often than not this is untapped potential. It has nothing to do with performance. It's not about what a person failed to do, but what that person could do—his or her individual capability to be great at something.

When you sincerely acknowledge a person's innate potential, even when there's little clear evidence of it on the surface, the person is likely to respond positively. Some people probably aren't even aware of their own potential, because all anyone ever focuses on is their visible performance, things that can be seen and quantified. On the other hand, you may know some people whose potential and achievements you've respected for years; but do they know it? They won't if you don't tell them.

Following this rule helps people value themselves. When they believe you really want them to live up to their highest potential and that you believe they can, they may learn to believe it enough themselves to actually make it happen. Help people reflect on their past accomplishments and examine their present strengths, and their performance may begin to match their hidden greatness.

Remember to look for my loving intentions.

Some readers may recoil at the "L" word in this rule. If you are among them, by all means substitute "good" or "honorable."

All people are presumed innocent until proven otherwise; don't suspect devious motives. The archaic management assertion that workers are prone to be lazy goof-offs just doesn't hold water. And there's no middle ground on this issue. When people feel you're not seeing their best intentions, by default you must be thinking they're up to no good; people sense this. When they think you don't trust them, they may come to feel unworthy of making a contribution. In time, they may become unwilling to assume any responsibility or give any significant effort to their work.

Like the second rule, this one is about listening, but at a much deeper level—it suggests going beyond an understanding of what people say to the desires behind what they say. Of course, you may not always find good intentions; there is corruption in the world, and there are some incurable malcontents in the workplace. This rule says you should look for good intentions, always assuming they are there.

Give people the benefit of the doubt. Even when someone has a performance problem, try to find reasons to praise the person. This helps get people on the right track when their performance is less than you expect. When you help people feel better, and then offer some insightful coaching, they'll have the emotional energy and responsiveness to take your advice and try again.

Tell me the truth with compassion.

From time to time, every manager must confront an employee about a problem of discipline, performance, or conduct. It's never fun, but there are ways of going about it that can make the experience extremely positive for both parties. Don't avoid such confrontations; but approach them with a combination of firmness and caring.

Always speak directly to the person you must confront, not to other concerned parties behind the person's back; the last thing you want is for this individual to hear about your feelings through the grapevine. Be respectful, never insulting, disdainful, or condescending. Even when a person has a performance problem, you should approach it in a way that preserves his or her dignity.

Talk about what you can do to help a poor performer improve. Don't just be the one who finds the problem, be a part of the solution too. Ask if there is anything you've done that has made his or her job more difficult. Start the discussion by reviewing the person's good qualities. Also, keep in mind that performance problems rarely require extensive modifications of a person's behavior; in most cases only a minor adjustment is necessary. Let the person know this; it makes them feel far less threatened by the discussion and more willing to change.

When someone is doing something really harmful to a coworker, a customer, or the whole organization, the person should be encouraged to change as quickly as possible. If a person is doing a generally good job but still has room for improvement, counsel them as best you can, and then let their own values motivate them. If people refuse to change on their own, you can take appropriate action, but always give them a chance.

You won't be able to predict people's reactions. Just be clear, compliment what you can, and address the situation honestly. When you tell people the truth with compassion, they realize that you appreciate them, despite whatever problem has arisen. It lets them know where they stand in the organization and removes doubt about how you see their performance. Once people know you value them and their contribution, they'll probably want to improve their performance as much as you want them to.

YOU MUST CARE

One of the most meaningful commentaries on caring and the creation of mutual trust came from an officer in the American military, the late General Melvin Zais. While addressing a group of field officers at the US Army War College, he made the following observations on what he believed was an officer's most important duty: caring about the troops.

> *You must care about your sailors, soldiers and your airmen. Now . . . all of you are saying to yourself . . . "Well, I care. What's this guy talking about?" Well, there are degrees of caring and there are degrees of personal sacrifice to reflect the amount of caring that you do . . . You're sitting out there wondering, "do I really care?" How do you know if you care . . . you care if you go in the mess hall. I don't mean with white gloves and rub dishes and pots and pans for fine dust. You care if you go to the mess hall and you notice that the scrambled eggs are in a big puddle of water and the 20 pounds of toast have been done in advance and it's all laying there hard and cold. The bacon is laying there dripping in the grease . . . the cold pots of coffee are sitting on the tables getting colder. If that really bothers you, if that really gripes you, you care.*
>
> *When I was in Vietnam, a quartermaster captain was bragging to me about the ice cream that they made at Camp Edmonds, and were taking out to the soldiers . . . I said . . . "How do the soldiers eat it?" I said, "You know they're all in little dugouts, and not all lined up in the mess hall, and they don't have mess kits out there and things like that." He said "I don't know sir." I said, "I know how they eat it. They pass that damn [gallon can]*

around and they stick their fingers in it and each one grabs some." I said, "Find some Dixie Cups and send them." He said, "Dixie Cups?" I said, "Yeah, Dixie Cups." Well, I use these little elementary things because I'm trying to illustrate a point.

When you're getting ready for the regular inspection and you know these guys are in the barracks, and you know they're working like hell and it's Sunday night, if you'll get out of your warm house and go down to the barracks and wander in to see them work and just sit on the foot locker. You don't have to tell them they're doing a great job. Just sit on the foot locker and talk to one or two soldiers and leave. They'll know that you know how hard they're working to make you look good. They'll know you care.

If you have a fine, uncommonly good-looking non-commissioned officer with muscles rippling down his chest, and a strong neck, and clean as a whistle, trim as he can be, shoulders back, the look of tigers in his eyes, and he says to you, "Captain, don't worry about it. I guarantee I'm going to take care of it." If you don't worry a little about that man, if you don't wonder what he means by I guarantee I'm going to take care of it, if you don't check to see if he's making his guys do push-ups until they're dizzy and sweating, if you don't wonder is this guy getting sadistic pleasure from pushing these kids around, if you don't make it your business to make it known throughout your office that you won't put up with that crap, then you don't care. But if that worries you, and you wonder and you go and check and ask questions and you make sure, you care.

You cannot expect a soldier to be a proud soldier if you humiliate him, you cannot expect him to be strong if you break him. You cannot ask for respect and obedience and willingness to assault hot landing zones ... if your soldier has not been treated with the respect and dignity which fosters unity and personal pride.

The line between firmness and harshness, between strong leadership and bullying, ... is a fine line. It is difficult to define, but those of us who are professionals, who have also accepted a career as leaders, must find that line ... I enjoin you to be ever alert to the pitfalls of too much authority, beware that you do not fall in the category of the little man with a little job, with a big head. ... I want to close by stating, if you care, I guarantee you a successful career. I won't guarantee that you'll be a General or Admiral, but I guarantee that you will improve your chances ten-fold. So, it is in your self-interest that you be happy in the devotion, love, and affection of your men. You will like yourself better. I sincerely believe that to be a successful leader in the idealistic sense, you must care.[3]

COACHING

The literature on business coaching is fairly consistent. Typically, coaching is used in conjunction with performance appraisals, to provide feedback on job performance, and point out where a person's skills should be improved. But coaching in the archetype sense has quite a different emphasis. While the conventional business coach is concerned with improving an employee's deficient performance, the archetype Coach emphasizes an employee's

---

[3]Lieutenant General Melvin Zais in a speech delivered to the Army War College, Carlisle Barracks, Pennsylvania, 1970.

positive, continuous development, based on training and disciplined practice. In the deepest sense of the word, a Coach is someone who maintains an ongoing, committed partnership with a player or team, demanding peak performance and supporting them in their quest to exceed previously set levels. See Figure 6.2 for a comparison of the conventional versus archetype role of a Coach.

On the subject of coaching, Tim Gallwey and Bob Kriegel, authors of *Inner Skiing* say, "Learning happens best when both instructor and student recognize that experience is the teacher. The

COACH	
CONVENTIONAL VIEW	**ARCHETYPE VIEW**
• ASSUMED TO BE AN IMPORTANT ROLE, BUT RARELY EFFECTIVELY PLAYED	• CRITICAL ROLE IN BUSINESS
• GROW WORK THROUGH PEOPLE	• GROW PEOPLE THROUGH WORK
• COACH SHOWS UP AT THEIR CLASSROOM WHEN CLASS BEGINS	• SHOWS UP AT THE BEGINNING OF THE PROJECT AND REMAINS FOR PRACTICE AND SKILL REPETITION
• AUXILIARY ROLE—OFTEN A TRAINER	• PRIMARY ROLE—OFTEN THE MANAGER
• PROVIDES INSTRUCTION BASED ON THE COACHES' TIME-TABLE AND READINESS	• PROVIDES COACHING BASED ON THE CHAMPIONS' TIME-TABLE AND READINESS
• ABSENT FROM PROJECT PROCESS INCLUDING END	• PRESENT FOR WINS AND LOSSES
• RESPONSIBILITY IS OFTEN A MATTER OF NEGOTIATION	• RESPONSIBILITY IS A PRIVILEGE FOR PLAYERS
• COACH IS EXPERT PLAYER	• COACH IS NOT EXPERT PLAYER
• DELIVERS INFORMATION TO PLAYERS	• GAINS INSIGHT INTO PLAYERS
• BOSS AS COACH SHARES THE CHAMPION ROLE	• COACH IS NOT THE CHAMPION, THIS IS NOT HIS FIGHT
• LOOKS FOR YOUR DEFICIENCIES	• LOOKS FOR YOUR HIDDEN TALENTS

Figure 6.2

role of the instructor is to guide the student into experiences appropriate to his or her stage of development. ... a teacher who likes to demonstrate how much more he knows than his student often reinforces the pupil's self-doubt."

As has been said many times in this book, fear is the great enemy; it can stop anyone in their tracks at any time. Effective coaching can quiet the mind and therefore quiet people's fear. Gallwey and Kriegel distinguish between two types of fear: FEAR 1, an immobilizing force, and FEAR 2, an energizing force. They describe the former as the "fear that is harmful because it interferes with our abilities to perform at our best, [and it] originates in the imagination ... FEAR 1 has a magnifying effect on our perception. When looking at danger, it greatly enlarges what it sees ... Our sense of competence is so depleted that we feel powerless to handle the situation confronting us."

"FEAR 2," Gallwey and Kriegel say, "is the body's natural response to challenge. It can exist simultaneously with courage, and often precedes the performances of athletes, actors, race-car drivers, soldiers or anyone in a high risk situation ... FEAR 2 focuses our attention in the present and lends us capabilities beyond our normal levels. Since this kind of fear is helpful to us, we need to learn not to resist it, nor to waste the energy it produces."[4] Good coaches help performers quiet FEAR 1 and embrace FEAR 2.

MARILYN REMEMBERS A COACH

Pondering the structure of the archetype for this book, considering Failure, Support, Celebration, and Impossible Dreams, I remembered an almost-forgotten experience from my work life. It involves someone who, for a while, served as my Coach. He was also my boss.

Our association lasted from 1974 to 1976, but it wasn't until after we'd parted company that I realized he'd been my Coach; we never saw our relationship as anything other than the normal boss-subordinate one. Had we agreed that beyond the conventional role he'd also be my Coach, I think we both would have grown much more.

---

[4]Gallwey and Kriegel, *Inner Skiing,* p. 58–63.

So why do I now see him as a Coach? Well, he took me under his wing. He coached me in ways that supported my growth and development, both as a person and as an employee. He was a stern task master, but I always knew I had his unconditional support. I'm not sure how I knew this; it was never explicitly stated, but he often said that I had a lot of potential and he wanted to see me grow.

We communicated openly and freely. He was an excellent listener and listened more than he spoke; he listened as a Mentor listens. When he did have something to say, it usually conveyed clarity and insight. Sometimes his words were painfully clear to me (he was excellent at telling me the truth with compassion). Other times, it took much repetition and some failures on my part before I was able to understand him. Often, he assigned me to project teams in roles that he knew perfectly well were far beyond my experience. When I protested, sometimes to the point of rebellion, he'd say, "I'm throwing you in the water knowing you can't swim. It's up to you to learn to swim; but I'll never let you drown."

This approach was hardly reassuring, and I didn't like it. I was anxious, nervous, and fearful I'd make a mistake. As a rule, I made many mistakes, more than I even realized. But true to his word, when I was in way over my head and thrashing about for air, he'd always pull me out.

It was at these rescue sessions that I'd learn the most, I was now open to learning. Having narrowly missed disaster, or so I thought, I'd be given a chance to catch my breath. He would then patiently have me review the chain of events with him and encourage me to probe for meaning and insight: what were the causes? What alternatives might I have considered? Sometimes the solutions seemed so obvious that I was truly embarrassed. Sometimes I felt as if I'd let him down, but he never said so. He was sometimes critical of my thinking process, but he always supported me and said he knew I could do whatever was required. When he listened empathetically my resolve and determination to try again were restored.

I was always amazed when he helped me find an elusive bug in some software I had written; I was a programmer-analyst, while he had never written a line of code in his life. Others who worked for him resented his lack of programming experience, feeling that a boss should know how to do all the jobs he supervised; this never bothered me. He'd sometimes walk over to my desk just before my frustration reached the breaking point, having been unable to find some pesky GOTO error.

He'd ask to see the code; that always seemed strange to me, since it must have been about as clear to him as Chinese. Thinking back on that now, it may have been just an effective delaying tactic, one that gave him a chance to think of what to do and time for me to calm down. He'd ask a few probing questions, and then we'd do a code walk-through; this was well before code walk-throughs had been invented. Sure enough, in less than an hour, the errors were found. As soon as he'd see the "Ah hah!" expression on my face, he'd disappear.

He was like that, always wandering around, poking in at just the right moment, and always knowing when to leave. His coaching style varied from gentle caring to brutally demanding; when I needed his help, I never knew what I'd face. His style seemed to vary with his moods. As I came to know him better, I realized that the pressure he was under weighed heavily on him.

During the time I worked for him, his own position in the organization deteriorated. Through several reorganizations, his position was eroded, and he lost responsibility and authority. He suffered a lot of pain from this personal defeat, but his support for me and the others under his care never wavered.

There were times when he needed our help more than we needed his, and we gave him unrestricted and unconditional support, returning what he'd given to us. I feel our support helped him personally, but he never regained the position he once held, and he never had a success we could all celebrate.

Luckily, he's an American and had a second chance. Soon after this time, he left that company and started over. Today, he has turned his situation around. He is a highly respected director of operations for a large, successful firm.

REACHING OUT

This boss acted as a demanding Coach every time he threw Marilyn into an assignment for which she didn't have much experience.

When he said he'd never let her get into trouble, she trusted him; but the boss would have made a colossal mistake if he'd assumed her trust was automatic. Over time, he had invested in building that trust. He'd demonstrated that he truly cared for her personal well-being, as well as her growth and development. Each investment in listening to her and showing that he cared reinforced her trust in him.

The trust of any person in any situation is earned over time; consider it as interest on investments in what Stephen R. Covey calls the "Emotional Bank Account." It's like the emotional bond that develops between Miyagi and Daniel. Making demands on people without a prior, carefully developed sense of trust will usually lead to disaster. Without trust, people have little faith that their managers will truly save them when the water rises over their heads. They tend to respond to criticism defensively, perform poorly, and perceive the demanding manager as brutal and insensitive. People cast emotionally adrift in such situations may carry out their responsibilities anyway, having no choice but to save themselves. This might make them tougher, like Army Special Forces troops, but at the cost of a growing attitude of cynicism and bitterness.

Covey, in *The Seven Habits of Highly Effective People*, defines an "Emotional Bank Account" as "a metaphor that describes the amount of trust that's been built up in a relationship. It's the feeling of safeness you have with another human being." Covey says deposits are made through "courtesy, kindness, honesty and keeping my commitments to you." He considers trust a resource to be fertilized and, at the right time, harvested: "I can call upon that trust many times if I need to. I can even make mistakes, and that . . . emotional reserve will compensate for it." But, Covey warns, "If I have a habit of showing discourtesy, disrespect, cutting you off, overreacting, ignoring you, becoming arbitrary, betraying your trust, threatening you or playing a little tin god in your life, eventually [the] emotional bank account is overdrawn. The trust level gets very low."[5] Our organizations are plagued by relationships in which trust has disappeared completely. Some emotional bank accounts were never opened, while others run for dangerously long periods in the red.

[5]Covey, *Seven Habits*, p. 189–190.

LEW LEARNS TO LISTEN

One summer a few years ago, in an effort to spend more time with my son, I suggested that maybe we should get together for 20 minutes or so once a week, just to talk. We could go up to his room, crank up the ceiling fan, and talk about whatever was on our minds. At the time, he was a senior in high school, working a summer job as a waiter's assistant. This was a fine job as far as I was concerned; I'd been very big on telling my son that everything in life is a learning experience.

We'd been getting along great and were enjoying each other's company—absolute bliss for a father. Before long, the 20 minutes of casual conversation became an hour and a half.

Mostly, I saw this as a great opportunity to pass some of my ideas on to my son, so he could learn from my experience; that's what fathers are for, right? But it wasn't until I began to listen to his ideas that things began to happen. While listening to stories about how he had solved a problem or handled a difficult situation at the restaurant, I began to find ideas I could use in my own job! Later on, I'd tell him how I had taken some of his ideas and actually used them successfully in the manufacturing operation at the Atlanta Works.

Using his ideas made it very clear that I was listening to him and that he was helping me in my work. It was then that I noticed a change in our conversations. It was a subtle difference, but tremendously important: he had started listening to me, really listening. A relationship of mutual trust and honest communication could not develop until I had demonstrated a clear willingness to pay attention to him, in a way that valued him and his ideas.

At home and on the job, people can live and work side by side for years without ever establishing a relationship of mutual trust. Sooner or later, someone has to open the door and display a willingness to care, even without compensation.

LEW COACHES FROM THE SIDELINES

At lunch during one of our Quality Forums, a production specialist brought up a problem that he felt helpless to resolve; he couldn't get along with a coworker in his area, even after years of working together. They had virtually no emotional relationship and weren't even close to being a team. In fact, they barely spoke to each other. I told him, "You are the difference you make, and if you want to make a difference, you can." I suggested he start by helping this guy do some things that needed to be done in his area.

A few weeks later, I bumped into the same production specialist, and he told me the following story:

After the forum, I decided to try what you suggested. So, I went out and got tags for this guy. I swept up his work area for him, and I emptied reels to support what he needed. When his machines broke down or he had wire breaks, I'd go over to help him string up. The whole week went by, and he never said anything, not even thank you. I was ready to give up, but I remembered you said it might take some time. So, I continued into the second week.

About three days into the second week, I started noticing tags showing up on my line. The guy was bringing over empty reels and sweeping up my floor. Before long, we started talking to one another. It wasn't long after that, we learned that we liked the same things. Our sons were going to the same college; we both liked fishing at the lake. We worked side by side every day for three years and never knew how much we had in common. We're now a team on the job and off. We work together and fish together.

Sometimes we need to give before we get.

TO DREAM . . .

Gallwey and Kriegel's concept of FEAR 2 reflects the flood of emotional energy that can be released by Phase I of the archetype. In the right environment, when people experience failure, they can become energized very much as Gallwey and Kriegel describe, their capabilities rising beyond what they thought was possible. In an environment where failures are outlawed and punitive reprisals are common, FEAR 1 takes hold of people. Mentoring and coaching help channel energy into FEAR 2, and thereby pull people into Phase II, where their energy and confidence can grow, and their competence grows to a higher degree of effectiveness. Once caught up in the channeled energy of Phase II, people are in a position to benefit from the last and perhaps most powerful element yet: the Impossible Dream. See Figure 6.3 for the contrast between the conventional and archetype views of the Impossible Dream.

A Crisis resulting from Failure creates frustration and discomfort but moves people forward. The Impossible Dream draws us forward. Both elements affect a change in the same direction, but the impetus for movement is different for each: one pushes, the other pulls. People are pushed from a failure but are pulled toward an Impossible Dream.

An Impossible Dream is something that people yearn for with all their heart, something they want so badly they can taste it. From

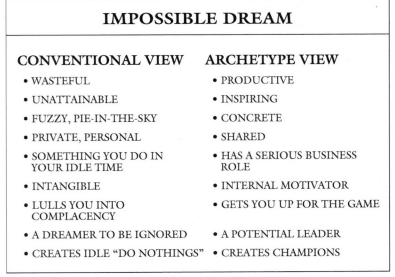

IMPOSSIBLE DREAM	
CONVENTIONAL VIEW	**ARCHETYPE VIEW**
• WASTEFUL	• PRODUCTIVE
• UNATTAINABLE	• INSPIRING
• FUZZY, PIE-IN-THE-SKY	• CONCRETE
• PRIVATE, PERSONAL	• SHARED
• SOMETHING YOU DO IN YOUR IDLE TIME	• HAS A SERIOUS BUSINESS ROLE
• INTANGIBLE	• INTERNAL MOTIVATOR
• LULLS YOU INTO COMPLACENCY	• GETS YOU UP FOR THE GAME
• A DREAMER TO BE IGNORED	• A POTENTIAL LEADER
• CREATES IDLE "DO NOTHINGS"	• CREATES CHAMPIONS

Figure 6.3

one's current vantage point, it may seem completely out of the question, so remote that there's no chance of achieving it. At the same time, it's so compelling that people sign up with their lives and prepare to give it everything they have. The Impossible Dream can be very personal or can embody the goals of entire teams and large organizations.

Such dreams can be difficult to express with words; as such, they should not be confused with mission statements. A dream is better expressed as a vision. It should be something people can envision, or imagine that they see, which depicts or describes how their world might appear, how it might be different, if the dream came true.

In his essay *The Power and Purpose of Vision*, Richard R. Broholm quotes Peter Senge, organizational theorist and MIT faculty member: "Vision is a picture of the future that you really care about. It is an expression of your core values, your sense of purpose ... It is a picture of the future that you really want to create. The problem with a mission statement is that it is very hard to get the sense of emotion, the sense of excitement. What makes vision different from mission statements and strategic objectives is that vision brings something from within and says, 'I really want to put my life energy into creating this.'"[6]

The last 200 years are full of dreams that have become realities. In 1807, the press ridiculed Robert Fulton's steamboat design, calling it Fulton's Folly, even after it had been successfully tested in Paris and New York. During the mid-19th century, people believed that trains could never go faster than 35 miles per hour, fearing that at higher speeds the air would become too thin to breathe. Before the discovery of anesthesia, even the most learned physicians were resigned to the idea that pain would forever be a necessary part of surgery. Before the Wright brothers historic flight in 1903, scientists claimed that all pursuit of heavier-than-air flying machines was a waste of time; some even offered technical proofs to explain why such machines could never succeed. The plain-paper copy process, upon which so many offices depend, was patented in 1939, but it took the inventor until 1950 to find a company that believed in the process enough to try building a prototype. That company was the

[6]Broholm, *Purpose of Vision.*

GATHERING SUPPORT **205**

Haloid Company of Rochester, New York. Ten years later, it changed its name to Xerox.

The vision of an Impossible Dream lies behind virtually all outstanding American business successes. In his book *Peak Performers,* Charles Garfield describes the power of a wonderful dream in various organizations. He recounts the story of Grumman Aerospace. A part of the *Apollo* space program in the 1960s, Grumman had won the contract for building the Lunar Excursion Module, the spacecraft that would actually touch down on the moon. The dream that fired these people was easy to see on any clear night, hanging in the sky. People had been dreaming about travel to the moon for thousands of years. At last some people were going to do it: they would be Americans, and they would be riding in a machine made by Grumman.

The Grumman managers knew the power of this dream, this vision, "to push back the frontiers of space and put Americans on the moon." They created slogans and posters and placed a model of the Lunar Module in the cafeteria. The resulting atmosphere was electric and contagious, for everyone became caught up in the dream. Garfield recalls that throughout the organization, people were striving "to find ways of doing outstanding work. To learn how to be a star in a big, complex organization. To be somebody."[7]

When Apple was close to being eaten alive by IBM, and working like mad to develop its Macintosh computer, President Steve Jobs moved his people to do the impossible with the power of a dream. He led his cadre of designers and software developers to extraordinary levels of achievement, exhorting them to see the uniqueness and distinction of their quest: "Opportunities like this don't come along very often. You know somehow it's the start of something . . . And it's being done by a bunch of people who are incredibly talented, but who in most organizations would be working three levels below the impact of the decisions they're making in this organization. It's one of those things you know won't last forever."[8]

One of the most beautiful, most inspiring dreams on record is that envisioned by Dr. Martin Luther King, on the steps of the Lincoln Memorial in 1963. We mention it here because today we

---

[7,8]Garfield, *Peak Performers,* p. 25 and p. 191–192.

have so few national dreams and we must keep this one alive. His was a dream of a future United States of America, which had left racial prejudice far behind:

> *"I have a dream that one day this nation will rise up and live out the true meaning of its creed; 'We hold these truths to be self-evident; that all men are created equal.' I have a dream that one day on the red hills of Georgia the sons of former slaves and the sons of former slave owners will be able to sit down together at the table of brotherhood ... I have a dream that my four little children will one day live in a nation where they will not be judged by the color of their skin but by the content of their character ... I have a dream today."*[9]

---

[9]Martin Luther King. "I Have A Dream." Speech delivered August 28, 1963, on the steps of the Lincoln Memorial in Washington, DC.

IDEAS FOR ACTION

PHASE II – SUPPORT

Don't be afraid to talk honestly about dreams. Don't convert the language of dreams to business verbiage.

– ◇ –

Create trend charts that display positive progress as an increasing curve rather than decreasing to zero. Remember Americans always want new challenges.

– ◇ –

Use language that energizes Americans. Use breakthrough rather than incremental improvement. Use quality challenges rather than quality alerts.

– ◇ –

Make time to discuss hopes, fears, ambitions, and dreams with team members and family members as a way to know each other. These talks help bond teams and families.

– ◇ –

Design projects for the family or team members so they are challenging and inspiring and have elements of the "impossible" built in so that when the project is complete the members shout, "Wow, we did it!"

– ◇ –

Value the people and they will value their work.

– ◇ –

Share your personal life at work and your work life at home.

Try something new, even if you never master it. Rather than getting hung up about your level of competence, just enjoy it and have fun doing it. Not everyone is a great wallyball player, but everyone can have fun playing. The point is to enjoy being a beginner again.

— ✧ —

The desire to explore new ideas, to have adventures, are typically American, and stimulate feelings of wondrous creativity. An important step in awakening your creativity is to rediscover that wonderful child in yourself. A child's day is filled with fascination, newness, and wonder. Rediscover the pleasure of creative fooling around. If you find it difficult, ask any five-year-old to teach you; they have a gift for tackling the impossible with zest, and repeated failure doesn't discourage them at all.

— ✧ —

Think in terms of doing things *for* people, not *to* them. In practice, this means letting people know what you expect, showing an interest in what they have done, supporting those things that are done well, and overlooking those noncritical things that are not done well.

— ✧ —

If you want people to contribute, find something for which to thank them. Each week, take the time to tell those who are important to you—employees, spouse, children—about things they have done that you appreciate.

— ✧ —

Visit informally with the people you supervise or parent. Don't look for problems, look for strengths—even in little things. Acknowledge all the positive things you have observed.

— ✧ —

Show your love. Actions can be as important as words.

Talk from the heart in the language of feelings and emotions. When you share in the accomplishments of coworkers, family, and friends, show them how you feel about them and what they have done.

$-\diamondsuit-$

Don't assume that you know another person's thoughts or feelings. It doesn't hurt to ask.

$-\diamondsuit-$

Help people rediscover their enthusiasm by encouraging them to go after their "impossible dreams." Encourage them to live each moment wholeheartedly, both the highs and the lows.

$-\diamondsuit-$

To encourage your coworkers, family, and friends to talk and share openly, put a lock on your negative reactions, criticisms, or judgments.

$-\diamondsuit-$

If you want to open the door to communications, rather than having it slammed shut, it may make more sense simply to listen.

$-\diamondsuit-$

Make the ordinary memorable. Look at everyday events with eyes open to the symbolism they hold as metaphors for caring. Team or family stories provide some of the most cherished memories and ignite the spirit.

$-\diamondsuit-$

Invest in future memories: A spontaneous visit to the folks in the shop just to say "hello" or a parent's decision on a lovely morning to forgo chores in favor of a bike ride with the kids. It's the composite of such moments when you take time for another person that creates an emotional bond. It's the kindness and caring all pasted together that form a beautiful memory collage.

When mentoring, become a possibility thinker and a hope generator.

– ✧ –

Allow people to own the result. It's staggering how far they will go if they do.

– ✧ –

Make requests in a positive rather than negative way. Your words should come out sounding more like invitations than put-downs.

– ✧ –

Punctuate your workday with bits of playfulness. Play jump starts internal batteries that have given out from driving ourselves too hard. Lightening up allows us to enjoy our day rather than merely endure it.

– ✧ –

Wonder at the world and dream questions bigger than yourself. Many of the world's artistic creations and scientific discoveries come from these ponderings.

– ✧ –

Let go and love. Children teach us about unconditional love. We love most deeply when we love with the full and indiscriminate compassion of a child.

– ✧ –

Don't stop learning. Never hesitate to seek guidance from family, teammates, and experts and then pass it on.

– ✧ –

Seek reinforcement for yourself. Everyone needs a mentor, someone to turn to for support. It could be a coworker, a member of the family, or a teammate.

Stay enthusiastic by discovering ways to revive your enthusiasm continuously. Brooding about bad times won't help. It's normal to get down; the important thing is not to stay down too long.

– ✧ –

Pay compliments. They express your awareness.

– ✧ –

Respect and enjoy the enthusiasm and excitement of others. There is something unforgettable about those times when you share in their enthusiasm.

– ✧ –

Share your own sense of excitement and wonder. Given the opportunity, others will eagerly share your interests.

– ✧ –

Make the time to share your dreams with others.

– ✧ –

Take the time to learn about the dreams of others.

– ✧ –

Rediscover the joy of hunting for those special "rainbows."

– ✧ –

Keep a journal describing four things that you want to remember about each day. You will become more observant and find that there are bright spots even on "bad" days.

– ✧ –

Support others by helping them remember what they already know.

Help others find out for themselves by asking questions and seeking to understand instead of giving answers.

– ✧ –

Don't box others in with one right answer. Creativity can be stimulated by letting the imagination roam.

– ✧ –

Sometimes communication with others can be improved by talking less and listening more. One-sided discussion closes another person's ears and mouth.

– ✧ –

Give specific reason's for your actions in present language. Don't fall back on clichés to justify your actions.

– ✧ –

You can learn a lot by asking two simple questions: what were the best times of your day? What were the worst?

– ✧ –

An excellent opportunity for prime-time communication is during team projects or family chores.

– ✧ –

Don't assume your family or team knows you care about them—show them. Begin with the smallest, easiest display of caring. You may be pleasantly surprised at the effects on the whole group.

– ✧ –

Pursue "gold-medal level" goals. Worthy and ambitious goals can excite and energize most individuals and organizations.

– ✧ –

Don't just say "We care about you." Find a way to prove it. Act immediately and personally in response to any problem that arises.

Words are powerful; choose them carefully. They can convey different meanings, e.g., customer versus partner, employee versus associate, hourly worker versus production associate, and judge versus coach. Focus on the language used in your organization; it conveys an important message.

– ✧ –

Remember to work with the whole individual—the mind and the emotions.

– ✧ –

When mentoring and coaching others, be supportive and understanding and allow the champions to choose their own way to succeed. Acknowledge their feelings without judging them.

– ✧ –

Use nonjudgmental language. For example, if two projects or tasks are different from each other, describe their differences; don't label one better than the other.

– ✧ –

Don't give up because a task is difficult or complex. Tell yourself it's no harder than challenges you have already overcome. It may simply require more time.

– ✧ –

The next time you take a walk, be open to the people around you. Smile and give them a pleasant greeting. Some real magic will happen.

– ✧ –

Try to find out a bit about what is going on in other fields. Breaking new territory and discovering more facets of yourself is what keeps you young, energetic, and curious about life.

Americans are like a rich father who wishes he knew how to give his sons the hardship that made him rich.

—Robert Frost

— ✧ —

Could a greater miracle take place than for us to look through each other's eyes for an instant?

—Henry David Thoreau

— ✧ —

I have learned that success is to be measured not so much by the position that one has reached in life as by the obstacles which he has overcome while trying to succeed.

—Booker T. Washington

7 EMBRACING CELEBRATION

TAKING TIME ALONG THE WAY

Emotions are strong components in Phases I and II. They provide the fuel needed to overcome the pain of failure, to learn new tools and techniques, and to battle against the forces of adversity. But there's nowhere in this quality process where emotions are more important than in Phase III. Figure 7.1 offers a comparison between the conventional and archetype views of emotions.

Phase III is called Celebration. To most people, this implies a single event that only occurs at the final completion of a project, when you achieve some sort of ultimate success or victory. It's easy to think of celebrating when a complex project is finished on time, under budget, and to the customer's lasting delight—all despite enormous difficulties, problems, and failures. Under circumstances like these, of course you would celebrate. But concepts like completion, project, and success should be viewed in a larger context, not just applied to big, one-of-a-kind undertakings. When a project appears hopeless, when an individual or team is struggling and their energies are nearly depleted, when frustration and anger are at their peak, some form of Celebration can change the dynamic; it can provide just the right antidote to break through the current impasse. These are the times for you to celebrate how far you've

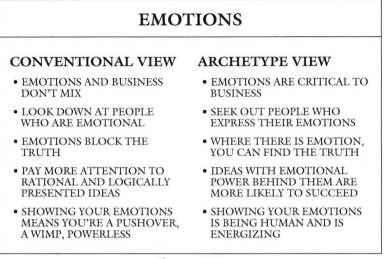

Figure 7.1

come. How close are you to completion? What milestone did you just pass in your project? How much nearer is your success? These are times to celebrate the people, their spirit, and their willingness to persevere.

LEARNING TO CELEBRATE

We hope you've found many ideas you'd like to try as you've read up to this point. However, if you're wondering where to start implementing the learnings from this study, it's right here. Begin with celebrations! When conducted according to the archetype definition, they serve to pull people into the other phases. For the contrast between the conventional and archetype views of Celebration, see Figure 7.2.

In Chapter 2, we stated that one of the Coach's jobs is to break down large projects into smaller, more easily managed activities. The Coach helps provide balance between tasks that are challenging enough to let people stretch their abilities but that are also finite enough to keep them from being overwhelmed. This balancing applies to Celebration as well. Managers are advised to stage lots of

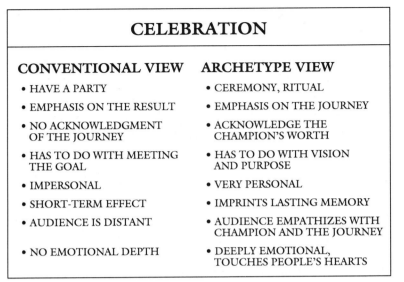

Figure 7.2

small, mini-Celebrations during the course of a project, each one appropriate to the events being celebrated, and to hold major Celebrations at major milestones. They should also encourage others to do the same—the more Celebrations an organization has, the healthier it will be.

The approach to any given Celebration should be matched to the event or success being celebrated. During a software development project, the correction of an elusive bug that only took three days might warrant a casual recounting of what happened during those three days, and maybe some coffee and doughnuts for the team. At a dinner party to recognize some people who won an important customer account after chasing it for six months, it's important to tell the detailed story of what happened during those six months.

The people involved should share the story of their struggle. The manager of such a team needs to encourage the members to describe their efforts at writing proposals, researching the customer's needs, and creating just the right graphics and presentation materials. In sharing their stories, the team members should describe their apprehension about the customer's response to their ideas and the nervousness they felt when the feedback began to come in.

If the story had a happy ending, great; but if not, sharing the story will help release the negative emotions that people feel. The experience will give them the energy to return to the drawing board, even more determined to succeed the next time. Effective, empathic listening, with no advice, is palliative; it lessens the intensity of people's reactions to a negative experience and lets them find their own solutions. The Celebration primes them for trying again; it's at times like these that people are most likely to reach out to ask for help from a Coach.

Simple and effective Celebrations can also occur during staff or project meetings. Try to conclude such meetings with brief success stories. You can introduce this idea very simply, by asking if anyone has had a success story to share since the last meeting. If you're met with puzzled silence, share one of your own. It could be an on-the-job story, about how someone helped you when you were late delivering something to the boss, or it could be a story about the pride you felt when your daughter succeeded in becoming editor of the high school paper. Wherever the story comes from, tell it from your heart. Include all the emotions you felt—both the highs and the lows. Emotions are good. They provide the fuel to propel us forward.

Over time, everyone will have stories to tell, many of which will be about other members of the group. As people grow more comfortable telling success stories, they will eventually take the greater risk of telling failure stories. Once again, as the leader, you may have to take the risk first, and you may have to do so more than once. If you can successfully encourage these people, through the vehicle of these stories, to make their failures public and concrete, you will have taken a tremendous step toward becoming a healthy, learning organization.

SHARING THE JOURNEY

In applying this archetype learning, we discovered a powerful new approach to Celebration, which we call a team renewal session. These sessions usually require several hours and are best conducted in a comfortable setting away from the office.

A skilled facilitator should guide and document the event. People close to the team, but not directly involved in its project, may be invited as an audience. Very gently, people should be encouraged to recall the beginning of the story and to describe what happened and how they felt at different points along the way. As their story unfolds, the facilitator can record both the specific events in the story, deadlines, meetings and so on, and the feelings that were associated with them. If the team describes the events and their associated feelings in enough detail, the participants and the audience experience a ceremony that approaches the richness and power of the journey itself.

The most remarkable characteristic of these Celebrations is that no matter how successful the people have been, they invariably choose to spend most of the available time talking about the pain of their experience. In many cases, they're still feeling pain, even though the project is finished and successful.

When the session is over, people should write down their personal feelings about the Celebration, being very specific about what it meant to them. This step offers the participants a final catharsis and supplies valuable feedback on the session for the facilitator. In the sessions that we have conducted along these lines, people consistently expressed a feeling that the team renewal session was the best recognition they had ever received.

We encourage everyone—managers, executives, supervisors, teachers, parents, and children—to experiment with this form of recognition. Making the deadlines, beating the odds, and winning are certainly important, but in and of themselves they do not motivate people to take on a new challenge. Inspiration, determination, and perseverance grow from hearing the personal story of someone's entire journey, in which 90 percent is struggle and only 10 percent is success.

THE WHOLE TRUTH

Phase III is about celebrating the worth of people—the Champions. See Figure 7.3 for the contrasts between the conventional and the archetype views of Champions. Many people have become so

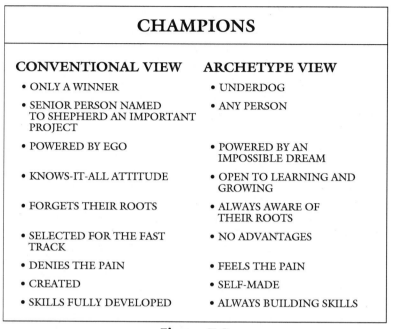

CHAMPIONS

CONVENTIONAL VIEW	ARCHETYPE VIEW
• ONLY A WINNER	• UNDERDOG
• SENIOR PERSON NAMED TO SHEPHERD AN IMPORTANT PROJECT	• ANY PERSON
• POWERED BY EGO	• POWERED BY AN IMPOSSIBLE DREAM
• KNOWS-IT-ALL ATTITUDE	• OPEN TO LEARNING AND GROWING
• FORGETS THEIR ROOTS	• ALWAYS AWARE OF THEIR ROOTS
• SELECTED FOR THE FAST TRACK	• NO ADVANTAGES
• DENIES THE PAIN	• FEELS THE PAIN
• CREATED	• SELF-MADE
• SKILLS FULLY DEVELOPED	• ALWAYS BUILDING SKILLS

Figure 7.3

caught up in the drive for achievement that the basic value of human beings has been lost. Too often and too easily, Americans come under fire for having lost the work ethic. Some critics are quick to label today's work force as lazy and uncaring, takers instead of givers. Such observations belie a serious lack of understanding. The American work ethic is as strong as ever, but it's often dormant, waiting to be tapped. Americans will never again be motivated as their parents and grandparents were—most of the social and economic conditions that motivated people 40 years ago have changed or disappeared. Sadly, contemporary American businesses have evolved attitudes and behaviors that repress the emotions needed to reignite the dormant work ethic. In the words of Kate Ludeman:

> *The work ethic—producing goods and services through long hours of hard work—is no longer serving America well ...*
> *The reason is that the work environment found in most American companies discourages people from working from*

the heart—from caring about the job they do and the people with whom they work. As a result, the work environment and the way we manage and define our jobs . . . discourages what we used to call labors of love.

In my experience as an organization psychologist and consultant, wherever I find employees who work from the heart, I also find managers who dedicate themselves to building up the self-worth of the employees. These managers tell the truth, share power, praise good performance and in other ways take practical action to show they care about their employees. These managers replace political maneuvering with genuine support for risk, innovation and growth. Their employees, free to put forth their best efforts, thrive in this environment and are able to create worthy products and services.

Managers I work with begin changing to what I call the worth ethic by thinking about their careers, and how they learned to be uncaring . . . Once these managers were able to understand the roots of their own uncaring attitudes, they began to break old patterns and to rebuild their own feelings of self-worth. They learned to care—first about themselves and then about others. And after they learned to care, they became emotionally ready to treat their fellow workers like family, no matter what their level was within the company. And when people work from the heart, productivity-gains result.[1]

As we made clear in the Celebration examples above, the most important element for implementing Phase III is sharing ourselves, by speaking and listening from the heart. In Kathryn Cramer's insightful book, *Staying on Top When Your World Turns Upside Down*, she says:

At the completion of a complex process . . . it is important for you to reflect on what actually happened and be able to tell the story. When you relive your growth process by recalling what happened and then expressing your reflections in the words to a story, you internalize your recollections and make them part of you forever. This is how you learn what went well and what

---

[1]Ludeman, "Embrace Your Workers."

turned out poorly. Your reflections reveal to you your own inner strengths that promote your accomplishments ... Telling and retelling your story, as often as necessary to fully release your negatively charged feelings, is one medicine that heals.[2]

She quotes a story from *Healthy Pleasures* by Robert Ornstein and David Sobel:

Thirty-three survivors of the Holocaust gave video-taped interviews about their experiences during World War II, while skin conductance and heart rate were monitored. Virtually all these survivors had suffered, many of them in silence, for decades. They had been displaced from their homes and forcibly relocated in ghettos. Many endured random beatings. Most witnessed the deaths of children, close friends, and family members.

Was it better for their health to disclose the most private aspects of their experiences? Those who more freely described their powerful trauma reported fewer health problems. Expressing both the facts of the trauma and the emotions seems to be critical for health improvement.[3]

Remember, regardless of how wonderful a project's final success might be, the archetype is lurking somewhere behind it; this means that at some point on the path to success, some form of failure occurred, and the people involved felt a lot of pain. Even if things finally worked out as planned and hoped for, it's a mistake to let this pain linger inside. That's why telling the whole story during a Celebration, recalling the worst moments as well as the best, does more than just energize people. It heals them, releasing the pain of initial failure and priming them for the next challenge.

The focus group session held with the Copper Shop team, described in Chapter 3, was an excellent example of Celebration through storytelling. At the time, those people had no idea they were celebrating, but the participants all reported feeling absolutely wonderful after it was over. They felt that something quite special,

---

[2], [3]Cramer, *Staying on Top*, p. 253–255 and p. 186.

almost magical, had occurred, linking them together and creating a feeling of power. After that simple round of storytelling, they were ready to take on any challenge anyone might throw at them.

ALL SHAPES AND SIZES

Celebrations and their associated storytelling do not have to be long or elaborate, nor do they need to involve medium to large groups of people. They don't have to follow headline-making achievements, and they don't even have to occur in an atmosphere of victory and success. Here are two stories, each giving an excellent example of Celebration, that fall outside the word's normal connotations. The first story unfolded shortly after an Atlanta Works employee named Joe was promoted to senior engineer.

Instead of taking the traditional approach of inviting this man into the boss' office to give him the good news, our management team decided to announce his promotion during a monthly birthday and service anniversary get-together. The celebration was held in the work area, where the people feel most comfortable and could openly express their genuine feelings and excitement about Joe's promotion.

Joe was unaware of our plan. He was completely taken by surprise and was moved by the sincere good wishes all his coworkers expressed. It became a special moment in his life, loaded with good feelings, and he was eager to share them with his family. Several weeks later, at a team get-together, Joe shared with me the wonderful way he told his family about his promotion.

In Joe's home, they have a special red plate. When any family member has had a particularly fine experience at work, school, or home, something about which the person feels particularly proud, the person removes the regular plate from their place at the dinner table and replaces it with this special red plate. The red plate is a signal to everyone else in the family. It lets them know that the person sitting in that spot has a story to tell; they all listen attentively, and join in the celebration of his or her success. This is a very simple, yet profoundly meaningful Celebration for all members of the family.

The next story is an outstanding example of the American talent for snatching victory from the jaws of defeat, a situation that began with what most people would say was a group with absolutely nothing to celebrate.

In 1977, Lou Holtz was coaching college football at the University of Arkansas, home of the legendary Arkansas Razorbacks. They were scheduled to play the Oklahoma Sooners in the Orange Bowl.

Two weeks before the game, Holtz discovered that three of his players, including two of his best running backs, had violated team rules; he had no choice but to suspend them. Another player was put out of action by an injury. The team's spirit was at an all-time low. The fateful game, the Miami Classic, was only days away, and each practice was an embarrassing exercise in bad timing, low endurance, and poor teamwork. Holtz called a team meeting.

He told them a story about a family tradition involving simple acknowledgment of someone's achievements. He said that whenever a family member did something well, everyone else would try to make the person feel special. This involved simple things, like letting the person pick the dinner menu and the topic of conversation at the table. "But the best part," he said, "comes when everybody in the room, one by one, tells that person something that is sincere and genuine. You say how much that person means to you, how much he has done to help you with your algebra, what a pretty dress she is wearing. You must be sincere."

And now Holtz told his team members to enact that very same ritual. First, a lineman stood up and complemented the Arkansas defense. Another player mentioned how important it was to have Steve Little on the team, the best field-goal kicker in America. An inexperienced running back, Roland Sales, who was sick and hadn't been practicing well, stood up and just said what a thrill it would be for him to play the game at the Orange Bowl with such a fine team. And so on.

Arkansas beat Oklahoma 31 to 6. Roland Sales set an Orange Bowl record with 205 yards rushing and scored two touchdowns.

With a simple, honest Celebration in the midst of failure, Lou Holtz made that team into a family and turned an underdog into a Champion: Oklahoma, ranked No. 2 in the country and confident of an easy win, didn't stand a chance. Lou Holtz had the American archetype inside him, alive and kicking; he also had the advantages of excellent intuition, a natural coaching talent, a lifetime of observing people's behavior, and an ability to energize his team. He knew the power of emotions.[4]

LEW RECEIVES A THANK YOU

It's amazing how very simple things can sometimes have an incredibly powerful effect on people. One day, I received a letter from a production specialist named Henry, who had attended one of our quality forums several weeks earlier. He had hardly said a word through the entire experience, and his letter explained why.

He began with deep thanks to me and Judy, the facilitator, for involving him in the forum and doing a great job running it. He went on to explain his silence during the session. It wasn't due to lack of spirit; quite the contrary. He was so overwhelmed by the discussion of quality issues and the environment of openness and respect, that every time he wanted to make a comment he would get too choked up to say anything. Now it was my turn to be overwhelmed.

A few weeks later, we had a visit from an executive vice president. With Henry's permission, I showed the V. P. the letter as evidence of how strongly our people felt about the quality effort. As a simple courtesy, I wrote Henry a short thank-you note for letting me use his letter.

Much later, I learned that Henry had reacted to my thank-you note as if it were an Olympic gold medal (and maybe for him it was). He posted it on the bulletin board in the shop. He showed it to his supervisor and told everyone how proud he was to receive it. He took it home to show to his family, and even his minister. Finally, he had it framed and hung it on the wall in his house.

At that point, I decided that if Henry could display my letter so proudly, then I was certainly within my rights to show off his. So, I had his original letter framed and hung it on the wall in my office; today, it's displayed proudly in my home. Amazing, what a simple thank you can do.

---

[4]Exum, "Football Story," p. 31–32.

ON RECOGNITION

Our discussion of Celebration wouldn't be complete without some comments on recognition and reward.

Too often in modern business, "recognition" consists of handing someone a token that has neither emotion nor any real connection with what the person did to be recognized. Such an item might be a plaque, a pen-and-pencil set, desk accessory, or other material gift, usually selected by the group or manager wishing to acknowledge a job well done. In the absence of anything else, people graciously accept these tokens of appreciation, but in their hearts they wish for more personal recognition.

In their efforts to systematize the recognition process, many well-intentioned managers have turned to outside companies—these firms sell a variety of gifts, beautifully displayed in glossy catalogs.

In these award programs, people who do the best jobs win points. When they've accumulated enough points they pick a gift from the catalog. The implied message is, "Want a better gift? Do better work." Some catalog-based award programs also include templates for letters of recognition and guidance on holding award ceremonies. These less material elements approach what our study data describe as the kind of recognition Americans want and need, but few of these companies sell their award package without the gift catalog.

People in the American workplace are starved for recognition. In the absence of anything else, they will accept gifts and prizes, but what can managers do to recognize people in ways that actually satisfy their emotional needs? In surveys and focus groups, the findings consistently show that what makes people feel recognized is being cared for, supported, and trusted. One employee summed it up perfectly: "What I really want more than anything else is for them to speak to me from their heart about how they feel about me and what I've done."

It's vitally important to understand that success is not a requirement for recognition or reward. People who are involved in personal risk, who have made an attempt followed by a failure, or who are working toward a success that's still months away are all prime candidates for recognition.

Conventional recognition ceremonies acknowledge only the achievement of success, the win, typically the last 5 or 10 percent of the whole experience. This approach has two big drawbacks. First, it forces the Champions to bottle up their memories of pain and failure, preventing them from experiencing the healthy and appropriate catharsis they deserve. Second, it holds the audience at a distance. A recognition ceremony that's exclusively positive makes the success seem too easy, faintly unreal, as if there were no struggle involved—people know in their hearts that this is not what they have experienced. At this type of Celebration, the audience may react with a mixture of envy, disbelief, and depression, because they feel so far away from succeeding and may believe they're not hearing the whole truth.

The right kind of recognition ceremony involves the sharing of emotional gifts, not cash, plaques, or trinkets. People receive gifts of sensitivity, empathy, and caring, items that have intrinsic value beyond anything on which you could put a price tag. They are gifts from the heart, which can stir people at a deeply emotional level. When people are touched in this way, their belief in themselves is rekindled, and their energy is revitalized. They feel the impossible is almost within reach.

Of course, there's nothing intrinsically wrong with material gifts. But if you want to give people gifts as part of a Celebration, the gifts should be something symbolic of their journey. When the Apple developers were ready to begin manufacturing the Macintosh computer, all the people who had worked on the project signed the mold for the plastic computer casing. To this day, every computer coming off the assembly line is a signed piece of work, bearing the names of all those whose creative talents gave birth to the new machine.[5]

As a general rule, the most powerful Celebrations are those that simply follow the archetype pattern for Celebration: supporting and acknowledging a team as the individual members relive their journey. The length of their journey doesn't matter—the entire experience may have been as short as two weeks or as long as two decades.

---

[5]Garfield, *Peak Performers,* p. 124.

THE NEXT CHALLENGE

If you have some lingering doubts about this recipe for success, think about the movies that Americans go to see over and over again. Would *Rocky* have been selected as best picture in 1979 if the film had told only of Rocky's victory? Would *The Karate Kid* have inspired other sequels if Daniel had beaten up the bad guys during the first 20 minutes? Would *It's a Wonderful Life* have become a national favorite if we hadn't witnessed George Bailey's pain and struggle, being driven to the brink of suicide? Deep in their hearts, people know that to win anything important takes a fight, a struggle, although they don't always behave that way. To have any sort of beneficial effect, reward and recognition must acknowledge this truth about life. When it does, people become more loyal to their team and gain energy to take on the next challenge.

IDEAS FOR ACTION

PHASE III – CELEBRATION

Share success stories often in any and every creative way. Take the last 15 minutes of every meeting to share success stories. Showcase people who have done something that they have never done before, no matter how small or insignificant it may appear to you, and refer to it as "doing the impossible." Encourage people to include the mistakes and failures.

– ◇ –

Give people the opportunity to share the positive feelings they have for one another.

– ◇ –

Recognition is a two-way street. Don't let rank get in the way. Giving praise and recognition ("PR") upward can encourage the behavior you find most supportive.

– ◇ –

Give recognition from the heart. If you're looking for a nice way to say you care, build a gift with your hands. It will be cherished and appreciated for years to come.

– ◇ –

Take time to express your appreciation in thoughtful thank-you notes. Few of us put the power of the written word to enough use. Receiving thank-you notes has a great impact.

– ◇ –

Challenge yourself to eliminate the negatives in dealing with other people. Put your effort into uncovering their positive contributions. Use the right-hand rule of quality: a simple handshake and "thank you." Say something complimentary to everyone with whom you live and work by the end of this week.

Collect reminders of special moments in time. Pictures of these occasions put into a photo album or scrapbook can serve as reminders. The Chemical Qual-A-Team (see Chapter 8) created a scrapbook filled with notes, flyers, purchase orders, team minutes, pictures, and name tags, each item representing a small step along their journey.

– ✧ –

Take the time to tell "remember when" stories. It is moments like these that draw teams, families, and friends especially close. "Remember when" stories may translate some painful moments into humor. It's the kind of experience when someone says, "We'll probably laugh about this in a couple of months." And the truth is, they probably will.

– ✧ –

Find symbolic ways to show how much you care. Recognition is symbolic and comes from the heart. Money alone doesn't do it.

– ✧ –

Make sure your gifts have special meaning for the recipient. Acknowledging and accepting what another person values can be a gift in itself.

– ✧ –

Pay attention to the casual comments made by team and family members. You will hear ideas for gifts that they may value without realizing it.

– ✧ –

Give your time or talent. Create a book of personal certificates to be redeemed for personal things, such as washing the car, ironing clothes, or baking a batch of brownies.

Don't wait for a special occasion. Whenever you give a spontaneous gift for no special reason, you're telling the person, more clearly than words ever could, "You matter to me."

– ◇ –

Give a gift that acknowledges and encourages a new stage in a person's life. It can be very reassuring.

– ◇ –

The easiest place to start using the archetype learning is with Celebration. Celebrate all successes, including the small ones and the honest attempts that failed.

– ◇ –

Make a conscious effort to lock in those special moments by bringing them to everyone's attention.

Anything one person can imagine, other people can make real.

—Jules Verne

— ✧ —

Hold fast to dreams
For if dreams die,
Life is a broken-winged bird
That cannot fly.

—Langston Hughes

— ✧ —

A rock pile ceases to be a rock pile the moment a single man contemplates it, bearing within him the image of a cathedral.

—Antoine de Saint-Exupery

8 THE IMPOSSIBLE DREAM

LEW'S PARALLEL UNIVERSE, PART 2

Throughout this book, you've experienced quite a mixed bag of information about the American quality archetype: study findings, examples from a variety of informed sources, and stories collected from our working and personal lives. In one way or another, virtually all of this material either led up to or expanded Marilyn's cultural study in 1986 and the series of AT&T archetype workshops that followed. While this material offers solid and valuable information on the American approach to quality, some readers may feel it's too academic, experimental, or uncontrollable to be of any real use. My advice to these people is to take a careful look at the recent histories of Xerox, Harley Davidson, Motorola, or any other American company that's undergone significant, rapid quality improvement during the last 10 or 20 years—I guarantee you'll see the archetype standing out like a beacon.

"Sure," you may persist, "it's real, but what can I do with it? It's an automatic, unconscious process. How do I put it to work? If it works once, how do I keep repeating the pattern—excellent questions. Even when you understand the archetype structure, how do you consciously put this understanding to work on the job? See Figure 8.1 for a visual description.

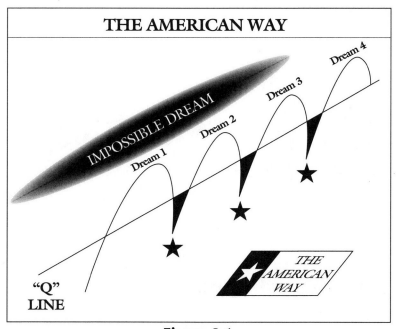

Figure 8.1

From 1984 through 1986, I had lived through a textbook example of the American quality archetype, driven by a crisis at the Copper Shop (as described in Chapter 3). During those experiences, I had been acting in ignorance, rolling along with the archetype process, without being able to see it. But since then, I have come to understand this process, and now use it every day. In one of my earliest experiences, I applied the learning on the job in the spring of 1988.

During May and June, I approached a challenging new management situation with a clear understanding of the archetype and its applications. I knew the signs that would let me recognize the archetype phases, and I knew what to expect at each critical point in the process. I was confident in my ability to respond to these signs in ways that would release people's energy, nurture their progress, and celebrate their success. And sure enough, when the dust had settled, these people had accomplished something that we all thought was virtually impossible.

Unlike the Copper Shop, this story started with a dream held close to the heart of one man in an organization of thousands.

A CULTURAL TRANSFORMATION

AT&T's Atlanta Works produces copper and fiber optic telephone cable and a certain amount of fiber optic hardware. As the story in Chapter 3 explains, this organization went through a major crisis during the mid-1980s due to a combination of ineffective management practices and new, aggressive competition. During a period of three years, the Copper Shop responded to this crisis with a greater focus on employee involvement (modeled after the natural work team concept) and a clear demonstration of upper management's commitment to quality and willingness to try new ideas. These efforts were successful—our financial picture improved, and our overall viability as a business began to look better and better.

In 1987, as a natural continuation of this process, the whole work force organized into different levels of quality teams: Quality Improvement Teams (QIT) for management and Quality of Work Life (QWL) teams for union employees, but these efforts were not particularly well integrated.

Going back as far as 1984, it's fair to say that the progress of our quality evolution was neither linear nor orderly; it proceeded through many organizational and personal initiatives that were seemingly unrelated but which were nonetheless all focused on the common goal of quality improvement.

In 1987, the Atlanta Works management adopted an organizational framework called the Quality/Productivity Architecture (QPA). This was an administrative umbrella, modeled after concepts developed in the Network Systems quality organization, and it appeared at just the right time.

Unfortunately, the QPA had something of a hidden booby trap: it had a wonderful emphasis on multifunctional teams and broad employee involvement, but it didn't specify a role for the union and it ignored QWL. Remember, the Atlanta Works is a union shop; at that time, about 2,000 of its 3,500 employees were members of the Communications Workers of America (CWA).

I led an implementation team, created to set the QPA in place and integrate it with our existing quality processes. Quality Improvement Teams were set up to support four different tiers in the organization: shop floor teams (small groups of supervisors and selected production specialists), line teams (department managers organized by lines of business), a plant team (a large management staff), and an executive team (just the highest level of management). As the QPA became active and expanded throughout the organization, the number of QITs began to grow very rapidly.

By the summer of 1987, the QPA's hidden flaw began to emerge. There were increasing numbers of conflicts between the new QITs and the established QWL teams; by the fall, there were four times as many QITs as QWLs. Neither management nor the union was happy with the situation. There was a lot of tension growing, so we called a time-out to explore alternative means of making the QPA operative.

By agreement between management and the local CWA representatives, a new QPA structure was established in October. This agreement eliminated the QITs and established the union as a regular participant in the executive team's weekly staff meetings; this was an important change for both parties. Through its attendance at these meetings, the union would have a voice in decisions concerned with running the Atlanta Works, while management would now have to be willing to air its dirty laundry in front of the union.

This agreement firmly established the QWL team as the operative problem-solving unit at the shop-floor level. This new QPA had a three-tiered structure: QWL, line teams, and executive teams. All QWL teams would now have defined roles for production specialists, union stewards, supervisors, engineers, and support people. Line teams would be associated with lines of business or functional support areas, and what had originally been the separate plant and executive teams were combined into a single executive team.

Making the union an active participant throughout our organization promoted mutual trust at every level. It was the beginning of a partnership: the union accepted its collaboration with management to achieve business-oriented goals as an essential element of job security; management, in turn, saw the support and participation of the union as critical to achieving organizational objectives. Throughout the Atlanta Works, people became involved. We had created an environment in which there was constancy of purpose, the opportunity to be creative and take risks, and the willingness to explore new relationships among employees.

Running parallel to our efforts with the QPA was a gentle groundswell: an expanding appreciation for the emotional dimension of quality. It resulted from a largely unheralded dissemination of knowledge about the American quality archetype through different areas of the Atlanta Works.

In December 1986, I had experienced Marilyn's archetype workshop, "Building Total Commitment to Quality," described in Chapter 4. During the next six months, more than 110 people from the Atlanta Works followed suit, including all members of the executive team, all department managers, and selected supervisors, engineers, QWL facilitators, and union officials. Many participants considered the workshop to be the best learning experience of their career. It fostered a sharper sense of team spirit and shared commitment. Relationships on the job became marked by more openness and a greater willingness to focus on the "why?" behind decisions. Caring became an integral part of the quality effort. It was symbolized by a heart-shaped button, worn by many Atlanta Works employees, proclaiming Jerry Butler's words, "CARE MAKES IT WORK." And Celebrations, large or small, became essential for showing support of the new culture.

While it was never explicitly stated, a sense of shared emotional values began to emerge, which resonated very positively with our QPA efforts. People began to see the new commitment to personal growth as an integral part of organizational growth. The evolution of our corporate culture now embraced more than the purely intellectual view—it also considered the perspectives of emotion, and the virtues of inspiration. This conscious change in culture created the environment that made the following story possible.

UNDERDOGS

The chemical purification department was small and was tucked in an annex building behind the two million square foot manufacturing facility. Besides being literally out back, they were also frequently out of sight and out of mind; a lot of people at the Atlanta Works didn't even know they existed. These people worked outside the manufacturing mainstream and had never been asked to start an official "team"; but the nature of their work (purifying chemicals used in the manufacture of optical fiber) and their relative isolation, had motivated the entire department to work as a team even before they became a QWL team.

For the first three months of 1987, the regular chemical department supervisor was out on disability, so the employees were left largely to their own devices. They lacked a formal structure, but they learned to work together with a flexible sort of interdependence; this strengthened their team spirit even more. They came to depend on each other for everything, developed a strong sense of trust, and often crossed the normal bounds of work assignments to help each other. There were also some advantages to their isolation: they had a degree of autonomy and relative freedom from company politics, which made it fairly easy to make changes and develop their own working style. When the supervisor returned from sick leave, they went back to their more conventional, if less flexible, functional roles.

In December 1987, the executive team instituted a plant-wide initiative to create teams in every manufacturing area, even the oft-forgotten chemical department. At first, the department supervisor resisted the idea, but other members of the operation were adamant and persisted until they'd been organized into an official QWL team. They called themselves the Chemical Qual-A-Team.

In groups of two or three, team members participated in two-day team-building sessions conducted for all QWL teams. This training stimulated some members to think about a project they could tackle, something of their very own. Nothing of significance developed; in fact, the team struggled for months just trying to design a team logo. Considering their vast potential and good intentions, most of their meetings were relatively unproductive.

BUY-IN-DAY: AN IMPOSSIBLE DREAM

Now we come to Mike Gravitt, a union production specialist and member of the Chemical Qual-A-Team. He was excited by the new employee involvement program and began looking for a way to spread its unifying energy throughout the entire plant. Inspired by a video of a quality-recognition day at another AT&T site, he conceived a modest proposal for a plant walk-through by labor and management to symbolize their coming together and their joint commitment to quality. Mike believed that no one in the United States, or the world for that matter, could outperform the Atlanta Works; the event he had in mind just might prove it.

Before Mike unveiled his idea to the whole team, he shared it with fellow team member Felton Davis, and together they explored its strengths and pitfalls. The idea still sounded good, but they didn't want to go through normal channels to get approval; they feared that if top management eventually said no, any intermediate managers who had said yes would be embarrassed. So they decided to bring their idea straight to the director of manufacturing. The response was favorable—he suggested that Mike and Felton pursue their idea as a team project, so they finally shared the idea with their whole team.

After a great deal of discussion, the team agreed that Mike's plant walk-through idea would certainly give the QWL program a shot in the arm and at the same time give the Chemical Qual-A-Team some much sought-after recognition. There was some concern about pulling off a project of this magnitude, but in the end they agreed to give it a try. This was the first time the entire team moved forward on anything of such significance. And coincidentally, within minutes they had all agreed on a team logo, something they'd been wrestling with for months.

Mike's dream was contagious and inspiring. The members of the Chemical Qual-A-Team were already excited about the QWL process and the other changes made to enhance employee involvement. They saw Mike's plant walk-through as a way to give quality teams throughout the Atlanta Works recognition for the excellent job they were doing and also to encourage more support for the QWL process. They began to call it QWL Buy-In-Day.

Mike pictured Buy-In-Day as a handshaking tour through the plant by management and labor, symbolizing togetherness and cooperation. It would be a celebration, demonstrating the caring and family spirit shared by the employees. There would be parades on each of three shifts, with music and costumes. Little did they know that, before it was over, Buy-In-Day would engage well over half of the total work force at the Atlanta Works.

While the team was still pursuing official permission to move ahead, a strange turn of fate worked in their favor. They had been invited to attend an organizational meeting to present the idea; but when the meeting time was changed, they were never notified. When they showed up at the designated conference room, the meeting was over. The only person still there was Bea Whelan, the operations manager. The team figured that an audience of one was better than nothing, so they pitched their idea anyway. She liked it, and what's more important, she told them she liked it. The team saw this as a form of recognition, their first, and felt wonderful just having someone listen to them. Bea promised to contact me, in my capacity as quality manager, to arrange a meeting. If I gave it a green light, they could move ahead and start working out the details.

I was swamped with work, and the last thing I wanted was another project. But Bea's description of their idea conveyed both excitement and a sense of mystery. I agreed to meet with the team members and hear them out.

THE PITCH

It was a hectic morning when they arrived; I was so busy I didn't even notice them walking into my office. I knew nothing about the Chemical Qual-A-Team but did recognize several of the people as former Copper Shop employees; this gave rise to a quick reunion. When the time came for them to make their pitch, they all began talking at once; their energy level was high, amazingly so, and they were all smiling big, genuine, winning smiles. After a minute or two they calmed down, and Mike told me about Buy-In-Day.

He started innocently enough, talking about this great idea to get everyone to buy into the quality effort, and then dropped a considerable bomb: "We'll just get everyone to form a team so that they can participate in this parade we want to have." Parade?

With that his words just poured out. He explained all about the band, the banners, the clowns; yes, clowns. He had a vision that everyone in the plant, all 3,500 or so, would either be in the parade or watching it. The parade would demonstrate unity between management and the union. It would symbolize their journey, a coming together from both sides, in contrast to the separation that's common between managers and union employees in so many factories around the country. Another team member explained the walking tour through the plant, planned for all three shifts, and a certificate that each participant would receive.

By now, I was regretting ever letting these folks into my office and was thinking of polite ways to say "forget it." I expressed concern about shutting the shop down for several hours on each shift and asked if there were any more details they wanted to mention. There were.

Now they talked about inviting all types of dignitaries: the CEO of AT&T, the Governor of Georgia, and President of the United States, to name a few; I did a double take, and they just kept going. All major TV stations would be invited to cover the event. College and high-school bands and cheerleaders could participate. They even suggested that the executive team dress up as cheerleaders; now that's what I call an Impossible Dream!

I had expected a very different meeting. I figured it would be the type where you hear a presentation and somebody shows viewgraphs. I wasn't prepared for this avalanche of ideas pouring out with a double helping of passion. The idea sounded crazy, but their spirit knocked me over and made me believe it might just work. There were a lot of questions to be answered. If I asked the executives to approve this, they'd think I'd lost my marbles; well, I guess many of them thought I already had.

Once we went ahead with this project, it would have to work— shutting down the operation would cost a bundle. But when I challenged the team, testing their resolve and commitment to the project, they never wavered. Their emotions made it clear that this was not just another team project; it was their very own Impossible Dream! In a flash of insight, as clear as if someone had held a big sign over their heads reading, "typical archetype behavior, circa 1987," I saw the American archetype alive and kicking inside these people. Their dream filled the office, spilled out the door, and made me utter a silent, personal "ah hah!" of realization.

The meeting lasted for almost two hours, but with all the excitement it seemed like minutes. As I began to visualize the possibilities, one concept kept coming back to me: everyone participates. In a plant of 3,500 or so, this was a very compelling thought; this Buy-In-Day might be just what the organization needed. Mike assured me that they were ready for a presentation to the executive team, which would be the next step to get final approval. I called the vice president of manufacturing and confirmed that within a couple of days the Chemical Qual-A-Team would have an opportunity to pitch their idea.

They had come into my office as underdogs. When they left almost two hours later, they had just won their first victory in a long series of challenges, most of which they had absolutely no idea were waiting for them. When I had tested their resolve and tried to shoot holes in their dream, they had summoned the strength needed to keep it alive. I found myself hoping they could pull it off again when they faced the executives.

THE NEXT CHALLENGE

No sooner had they conquered one challenge, me, then they were faced with another. It threw them into a considerable panic. They had assured me that they were prepared to meet with the executive team, but the reality was they didn't have all the details worked out, especially regarding that touchy business of shutting down the plant. And they had never before made a formal presentation; you know, the kind with an agenda, goals, objectives, expected results, and, of course, viewgraphs.

The Chemical Qual-A-Team got lucky. Its presentation to the executive team was rescheduled, giving the team a week and a half to prepare. Lisa Green, a team member who had recently completed some computer training, helped them create the necessary presentation materials. With her coaching skills, they organized their rather sketchy idea into something that came close to a manageable project.

When the time came, Mike made the pitch to the executives and received approval in less than an hour—his presentation was an unqualified success. But afterward, their individual looks of surprise expressed a clear, common thought: Now what? How will they get started? Who will help them? Who will approve the details? Was this a bad idea? During a period of several days, all the people on the Chemical Qual-A-Team sank into a panicky, uncomfortable state: a textbook archetype Crisis. They had overcome the first two obstacles, but now faced an even tougher one: making good on their enthusiastic claims. Seeing the familiar signs of the archetype, I once again felt that silent "ah hah!" of realization. I assured the team members that they could do the impossible—that in fact, they were in exactly the right frame of mind to get started. I assured them that my door would always be open and encouraged them to talk to me if they ran into trouble.

The next day, I met with several members of the team to talk about Buy-In-Day and was shocked to learn how little detail they actually had to support their plan. They had spent the last week and a half just getting ready for the presentation, spending hours in preparation and practice, but had not worked on the event itself. The presentation alone had taken more time than they'd thought it would. They had done a lot of it on their own time, because they still had an operation to keep running. I encouraged them to set the next phase of the project in motion: public relations. The Chemical Qual-A-Team had to spread the word to other teams and organizations to get their firm commitment and support.

The team scheduled Buy-In-Day for Friday, June 24th, the day before another Atlanta Works event, the Family Outing called "Flying High with Family Pride." The outing was a way for AT&T to let employees know that it cared about them, even beyond the bounds of the Atlanta Works parking lot. Choosing this date established a fixed time frame in which the project had to be completed, roughly 60 days. This tight schedule created another crisis; it generated a sense of urgency and produced the energy needed for the team to move into action quickly. After that, it was a matter of a lot of hard work, fueled by a dream. This is somewhat reminiscent of the situation in *The Karate Kid,* after Miyagi set up the tournament as Daniel's Lawgiver.

Slowly, the members of the Chemical Qual-A-Team realized what they were facing. During the next two months, the needs of the project would clearly require huge amounts of overtime. The team worried about this; they didn't want the project to look like some kind of scam, contrived as a way to garner them some extra pay. Their solution was a simple agreement among the chemical department employees, based on the interdependent working style they'd developed in the past. When a person was absent from regular work because of attending to some Buy-In-Day business, the others would simply fill in and pick up the slack. This was the case no matter what needed to be done or whose job it was: production specialist, salaried associate or engineer. If there was work to do, they did it—without overtime pay.

This agreement strengthened their emotional bond. It also provided the basis for a new way of thinking, a way of approaching challenges with an attitude of creativity, mutual support, and a belief in limitless possibilities. Most of the team members realized they had never done anything like this in their lives. They agreed that when problems arose, they would sit down together and work them out. They agreed to hang on no matter what and never to dwell on losses or setbacks, lest they become discouraged and defeated. During the next eight weeks they would be confronted with many reasons to throw in the towel, but their will always carried them through. They didn't know how or when to quit. As it turned out, they were just ordinary people with an extraordinary amount of determination.

INTO THE FIRE

The next four weeks were hectic, to put it mildly. Members of the Chemical Qual-A-Team gave six to eight presentations a day all over the plant, selling the idea to other groups and teams. They pitched it to QWL teams, line teams, and support groups, to audiences as small as three and as large as 50. They gave presentations on all three shifts, which meant some of them had to cope with long hours, an insane work schedule, or both. Just putting the word out about Buy-In-Day was a monumental task; but getting people to really buy-in, to commit time, resources, and energy, was even tougher.

It was during this period that the efforts of Lisa Green, the engineer on the Chemical Qual-A-Team, proved invaluable. Like me, Lisa had been through Marilyn's workshop on the quality archetype and was therefore in a great position to offer support at the right time. Our shared knowledge of the archetype gave us a certain heightened perception of the team's activities; we observed their behavior and saw the deeper, unconscious archetype structure. Lisa and I ended up working very closely for the duration of this project and leaned heavily on each other to help keep the team going. We served as Mentors and Coaches, providing the team members with support and helping them develop the necessary skills.

During this phase of the project, Bea made a point of mentioning the team's efforts in meetings; it was her way of giving them recognition and support. Ordinarily, this would have been a fine thing and would have made the team members feel great. But Bea kept referring to the team by the name of their supervisor. It was always "her team" and not "the Chemical Qual-A-Team." Along the same lines, when Bea wanted to contact the team, she would send messages through their supervisor, not through the Chemical Qual-A-Team chairperson. This hurt the team members very deeply.

A lot of resentment began to build, creating an emotional time bomb that threatened to destroy the team's energy. The members were becoming angry. They felt their supervisor was getting all the credit while they were doing all the work. Bea meant well and was simply going through normal channels. But without knowing it, she was hurting the very people she wanted to support and jeopardizing the entire project. The team felt that Bea's habit of contacting their supervisor, instead of going directly to them, reflected how little management really supported the new QPA structure; it had been designed specifically to eliminate this sort of conventional business hierarchy. Bea's actions implied an attitude of make-believe empowerment, form without substance. The team saw this as management's desire for absolute control, even at the expense of their morale and the success of the QPA.

This situation offers a valuable lesson for managers. If you accept the importance of emotion in the workplace, one of the chief concerns of this book, then you must stay informed and aware at a deeper level than traditional business wisdom suggests. You must always know how people feel. Never take it for granted that things are okay—prove it to yourself.

DOWN TO THE "NITTY-GRITTY"

While still working hard on selling its idea, the team now began to determine the actual shape of the event. As more and more people became interested, the project began to get out of hand, and before long there were more than 200 people directly involved. Taking advantage of the QPA's freedom and flexibility, they established assistant steering teams on the off-shifts to duplicate the functions of the primary Buy-In-Day team. Members of the Chemical Qual-A-Team met with these "Qual-A-Team Irregulars" every two or three days to exchange information, to answer questions, and to coordinate the effort. The commitment of time needed to support all this activity became enormous.

The team turned to other steering committees to look after the project's rapidly multiplying details. Each member worked directly with at least one steering committee and facilitated the necessary communications and hand-off activity. There were committees to look after the advertising, the walk route, audiovisual support, props, T-shirts, caps, and a large collection of other needs. The chairpersons of each committee had to meet at least several times a week—the activity was so robust that it was difficult to keep up with all the changes. When team members felt overwhelmed, Lisa and I would encourage them. We would listen to their concerns, help them clear away roadblocks, and, most important, help reaffirm their worth and the value of what they were doing.

There were many ways in which I tried to show encouragement. When I met a team member in the hall and talked for a few minutes, before walking away I would always say, "Remember to keep on keeping on." I tried to show up unexpectedly for their presentations and meetings, as much as my schedule would allow. I often called them Champions and would applaud or offer some other recognition for their accomplishments, no matter how small. After a successful presentation, I would always clap my hands and tell them it was time to celebrate.

There were many setbacks. Each day, the team would come to work and wonder what else could possibly go wrong. One member commented, "It feels like the seven of us are against 3,500." At one point their efforts actually came to a dead stop—very scary. They had reached an impasse and simply could not get anything else done. Most believed that the problem was management's attitude toward the project. They felt too much of their time was being wasted talking about the roadblocks, and this didn't leave enough time to work on the project. In a bold and intelligent move, very much in the spirit of the QPA, they sat down with the appropriate management people and got it all out on the table. After talking things over for three hours, they reached a more or less satisfactory accord.

This was a great victory, and did wonders for their morale. Our coaching and mentoring was working—the team members were gaining confidence in their ability to handle crises. One member said, "I'm beginning to believe in order to make miracles happen, we need people who make it tough." The seven members of the team became even more firmly committed to the tasks ahead. The conflict had united them.

Again and again, various people told the team that the project could not be done, that the task was too big for them. This just fueled their determination even more. They hung on like limpets, completely unshakable. Whenever a member of the team was backed into a corner, their united strength would surface, and they would rally with a spirit they never knew they had. They became Champions, determined to win and do the impossible. They also became bearers of hope and excitement for people not directly involved in the project.

EXECUTIVE ACTION

It was pretty clear from the start, to me at any rate, that when the executive team approved the project, they really had no idea what they had started. The scope of the project had been poorly defined, but they hadn't challenged it. Basically, they said it sounded like a very nice project, so why don't you go ahead. No one from the executive team had been assigned to work directly with the team, and suddenly this tremendous flurry of activity was attracting a lot of attention—the executives were getting concerned.

They asked Bea to step in to oversee the project, but it was too late for her to take any real control; things had reached what could be called epidemic proportions. She raised lots of questions about minor details, like the color of the Buy-In-Day T-shirts, who should receive them, and whether or not they should also purchase hats. The team saw this as interference not support. They were already working 12 to 14 hours a day and couldn't afford the time to rehash decisions with Bea. They had made commitments to purchase certain items, to be supplied according to certain specifications, but to satisfy management it looked like they'd have to change some of the orders.

The short lead time, and restrictions on their token budget, made this turn of events extremely difficult; once again, they turned to Lisa and me for help. We responded with more empathic mentoring. We gave additional encouragement and support, now delivered with more energy and enthusiasm. I continued my coaching efforts with a crash course in executive politics; because Bea was now part of the process, everyone would just have to learn how to work with her to satisfy the executive team.

These combined efforts were successful. What had threatened to become destructive interference eventually subsided into the background.

THE CONTEST

While the original concept of Buy-In-Day had been just a parade or walk through the factory, the idea of team floats emerged as the details developed; this idea finally gave rise to a float contest. A float competition would give the teams their own Buy-In-Day project. It would help unite the different teams and focus their energy on the event.

The contest theme would focus on why the teams believed in the importance of quality and the QWL process. As it turned out, this contest was just what the project needed to get people fired up. They liked the idea of competing. Their competitive spirit drove the project's momentum like adding fuel to a fire. The float contest was something that everyone could work on side by side as equals, as there would be no experts.

It didn't take long for the floats to become the focus of Buy-In-Day. When offered the chance to create something that had never existed before, the people became inspired. They were challenged to be the best, and they loved it.

Concerned that some employees might become preoccupied with building floats, the team encouraged people to build them off the premises or after hours. They also created rules that stipulated that no company funds could be used for float construction and limited the amount of time that could be spent on such activity during normal work hours.

It was amazing how people contrived to build their floats in great secrecy. Teams found hiding places throughout the factory and kept their floats under lock and key. Many built their floats in home garages and only brought them to work on the day of the parade. The members of the Chemical Qual-A-Team hoped that their coworkers would make a personal commitment, just as they had, and build the floats on their own time or swap out their team meetings to gain time for the project. Many people did just that.

While only a limited number of teams could win awards in the float contest, the Chemical Qual-A-Team wanted each participant to end the day feeling like a winner; they prepared a beautiful certificate that they would present to everyone who participated as a symbol of their journey. It read, "I bought into QWL." They also created six float award categories: first place, originality and creativity, best realization of contest theme, best color scheme, best team spirit, and most organized team. With three shifts in the plant, this meant a total of 18 awards. Each team would be photographed, and the pictures would be framed and displayed in the learning center. Other team photos, taken weeks before the parade, were displayed in the work area as a way to stimulate pride among the teams.

The last two weeks before Buy-In-Day were all-consuming for the Chemical Qual-A-Team. During this period, they all worked 12 to 15 hours a day and grabbed their lunches on the fly. During the last week, most team members put in 20-hour days. The consumption of coffee that week set records that stand to this day.

AT DAWN WE MARCH

Finally it was June 24th: Buy-In-Day. People began to arrive around 4:00 AM; the first parade was set to start at 4:30 AM. Several hundred people gathered at each of the staging areas, ready to march through the factory at the designated moment.

The route for the plant walk-through had been designed to wind through offices and shop operations, finally arriving at something called the "Bridge of Hope." This specially designed T-shaped bridge symbolized Mike's dream of management and the union coming together and staying together. The plan called for the union president and the manufacturing vice president to enter the bridge at the top of the T, from opposite sides, and then walk together down the trunk of the T, symbolizing the new path they'd carved out together. Then the real parade would start—the teams with their floats, costumes, and banners would march down the main aisle of the plant, accompanied by a live marching band.

As 4:30 approached, the emotional climate grew a little tense. Rows of floats, representing various interpretations of the quality theme, had been neatly lined up and were ready to go. Some people experimented nervously with a little marching-in-place. Most hadn't marched in a parade since childhood and were feeling slightly awkward at the prospect of striding and waving through their familiar workplace in front of bosses and coworkers. I could hear chuckling from a few clusters of people. Some were sharing stories about scouting or high-school band experiences, and everyone was smiling. A group of night shift production employees at a nearby staging area started waving to us; we waved back. Some marchers had dressed like clowns, but most wore their special Buy-In-Day blue shirt and hat, with the logo created just for this event. The quiet laughter built up to a roar. When the signal to start finally came, all feelings of foolishness dropped away, and it became pure and simple fun. The awkwardness was replaced by a tremendous rush of excitement.

One team member guided the union marchers, while Mike Gravitt guided the management team. The route was well marked, but he still walked in front to set the pace. With a walkie-talkie, he kept in touch with the union guide who had started from the opposite side of the factory. Finally, the two segments of the parade came together and crossed the Bridge of Hope. When the CWA union president and the Atlanta Works manufacturing vice president met on the bridge, they shook hands and exchanged a quick hug—everyone cheered, waved flags, and blew horns; then the real parade began.

Smiling marchers crossed the bridge, shook hands, and walked together down the factory's main aisle. The teams assembled and marched down the aisle with their floats—the timing and precision of all this activity was unbelievable. As people marched past the judges, the float builders showed their trump cards: cannons fired and belched smoke, gadgets turned and produced splashes of color, and people twirled on mechanical turntables and danced elaborately choreographed routines. A sea of blue caps and blue shirts poured down the main aisle, with an assortment of flags, floats, and banners. In total, there were more than 60 floats for the day's events, each one a masterpiece. One float, a management brainchild, was a casket containing the old management style, intended for burial. Another float took the form of two old men with long, gray beards being pushed around in separate wheelchairs. A banner hanging from the top of the chairs read, "You're Never Too Old to Change."

After the parade, we all moved to the cafeteria to share in the festivities. Everyone agreed it was a fantastic event. You didn't need to ask anyone—you could see it in their faces. People talked about the great job everyone did, how beautifully the event had been organized, and how much pride people had put into their floats. Everyone I talked with had been powerfully impressed and without question felt like a million dollars.

The event was repeated twice more that day, each time with the same level of energy and excitement. By the end of the day, most of the union officials and management team members had been awake almost 18 hours and most people on the Chemical Qual-A-Team had been up almost 48 hours, but they all left feeling fulfilled and carrying memories that would last a lifetime.

What had started as a modest team-building idea had become an all-day festival. On company time, more than 2,000 employees from all ranks of the organization whooped it up to show their support for quality. On all three shifts, feelings ran high as people celebrated the building of a bridge between the different segments of the Atlanta Works. These people had achieved a real triumph of the spirit, both human and American.

Almost 60 days of superhuman effort had culminated in one day of celebration, and then it was all over. During those 60 days, the place had been alive with inspired energy. In the process of organizing the event, the Chemical Qual-A-Team had developed technical, organizational, and negotiating skills they never dreamed they would possess.

LOOKING THROUGH THE ARCHETYPE LENS

What Mike Gravitt and his teammates accomplished in eight weeks was nothing short of a miracle. But there are many people like him in businesses all over America, individuals and teams fighting for their dreams.

The Chemical Qual-A-Team started out with a tremendous advantage that Lisa and I really understood: they were underdogs. In fact, they were made-to-order underdogs in just about every conceivable way. They were physically isolated from the main manufacturing facility, literally outsiders, working in a small, unimportant-looking building. This meant they were also strangers. They were an unknown group in a work environment of 3,500; most people didn't even know they existed as a team, let alone as individuals. Separate, isolated, and ignored, they dreamed of doing something great for the Atlanta Works, something that would let them be recognized. Buy-In-Day would accomplish this, putting them on the map in a big way. This is the astonishing hidden power of underdogs—their desire to win is far greater than that of people who are winning already.

While a few members of the chemical purification department had worked together for years, the Chemical Qual-A-Team had only come together one year before Buy-In-Day. After the inauguration of the QPA, and about six months before Buy-In-Day, the group went through two days of team training. This experience inspired them and was probably an important step in preparing them to pursue their dream.

Once they took on the project, their energy levels increased 1,000 percent. For months, the team had been searching for an identity, something to make them stand out in the organization—but without an Impossible Dream, even the task of choosing a team logo was a major undertaking. But Buy-In-Day changed all that. Realizing they needed to make presentations, post flyers, and manage hundreds of other details, they chose a logo in less than an hour. The Impossible Dream of their task inspired them, and the crisis of being under such pressure released their energy.

They knew very little about each other and didn't know what to expect from fellow team members. Suddenly, they were engulfed in a major undertaking. The odds against their succeeding with Buy-In-Day seemed enormous, but as they got into the project, they refused to believe it couldn't be done. They began their quest with a dream and would complete it through a lot of hard work and an unshakable determination.

Some members of the organization didn't actively support Buy-In-Day, which led to a difficult emotional crisis. At first it hurt the project and the team, but in time this conflict helped bond the seven team members who would eventually do most of the work. They became people with a common goal, whose lives were ignited by a meaningful purpose. They turned their enthusiasm and their faith into an asset. Each time they were told the project was impossible, their resolve grew even stronger.

PHASE I: CRISIS AND FAILURE

The team lived the American quality archetype.

Their project was born out of a dream and fueled by a Crisis, self-imposed though it was. First, the team truly believed this project had to be a winner; if it failed, it could conceivably weaken QWL and undermine the years of hard work that had brought employee involvement to its current high level. Second, their own reputation was at stake; there were many "outsiders" who thought the project was a waste of time, never wanted it approved, and didn't support it when it was. Some of these people actually hoped it would fail.

The team had to fight this resistance. Frequently, they were criticized for their lack of experience managing a major project. Many times, they were told it couldn't be done, but this just inspired them to work harder. Much of their time and energy was spent dealing with a certain lack of support within the organization. They made many attempts to engage all their members, but they failed. And finally, when things started rolling, management jumped in and tried to take control. Many mistakes were made along the way. Throughout this period, Mike was the Lawgiver. He never for a moment gave up on his dream, and he pressed his team members to keep going. Figure 8.2 illustrates Phase I of the process.

PHASE II: SUPPORT

The team had two Mentors: team member Lisa Green and me. We understood the nurturing process and the importance of support during this stressful period. The toughest thing we had to do in the 60 days before Buy-In-Day was convince the team members to have the courage to let their dream become a reality. As Mentors, we listened to their pain, assured them it was only human to make mistakes, and constantly reempowered them. Team members came to us when they needed someone to nod approvingly, had to unload their frustrations, or lacked confidence. We also gave them reassurance, removed occasional barriers, offered shoulders to cry on, and helped them renew their will to continue. Frequently, Lisa would say, "I keep telling them, 'you can do it.' There's no telling what will happen when they believe me." In addition to mentoring, Lisa provided most of the coaching. She coached them in organizational and presentation skills, and whatever else was needed to make the project possible. Figure 8.3 illustrates this part of the process.

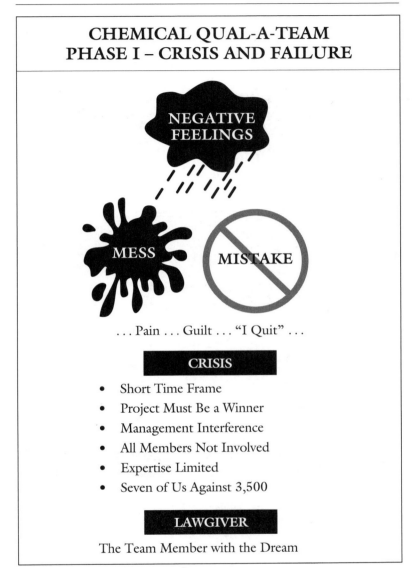

Figure 8.2

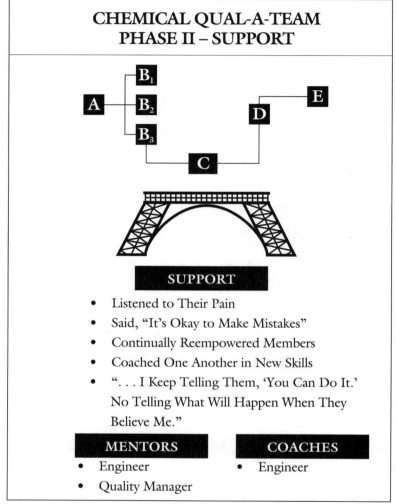

Figure 8.3

PHASE III: CELEBRATION

For the Chemical Qual-A-Team, Buy-In-Day fulfilled an Impossible Dream. In a highly emotional fashion, Buy-In-Day symbolically recreated the union and management's journey toward unified commitment to quality. It reaffirmed the worth of the QWL process and ensured its continuance. It also revealed the creative energy of the individuals and teams of the Atlanta Works. For many, it was the first time they had witnessed the potential of our people and the power of real employee involvement.

The story was an inspiration to many. The team did more than unify the plant around QWL; they also inspired other people to do things they didn't think were possible. David Nottingham, a production specialist, expressed it this way: "Buy-In-Day was good for me because I was able to reach down and find energies in myself that I never knew I had." The team had achieved a strong sense of accomplishment and cohesion. This part of the process is illustrated in Figure 8.4.

Several weeks after Buy-In-Day, Marilyn and I conducted a team renewal session, an archetype Celebration with the Chemical Qual-A-Team. This session, and others like it, were part of the cultural change taking place at the Atlanta factory. We wanted the team members to relive their success story. This would help us learn what worked and what needed fixing in the QPA, gain a better understanding of the team's success, and we hoped, find ways to reproduce it. The session also served to recognize and refresh the team.

We opened the session by asking the team members what they had gained from the experience. Without exception, they cited dramatic personal growth and expanded levels of self-confidence; they now had an unshakable faith in their ability to get things done. They were also pleased that they were able to use the QPA structure itself to win support for its unanimous acceptance. The members of the Chemical Qual-A-Team had become true champions and were recognized as such by most, if not all of the people at the Atlanta Works.

During the six-hour session, the team members talked about the events that led up to the parade and the few weeks following it. For the full six hours, they talked about the pain, the problems, and the barriers, not the successes and victories. Even though Buy-In-Day had been an unqualified success, many of them were still preoccupied with pain and were unable to let go of it. They didn't talk about the festivities or the wonderful things they had achieved. They only talked about what didn't go right.

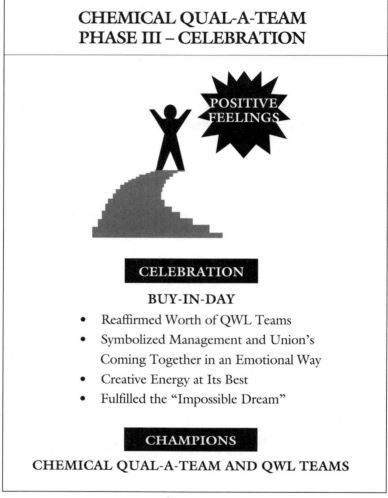

**CHEMICAL QUAL-A-TEAM
PHASE III – CELEBRATION**

POSITIVE FEELINGS

CELEBRATION

BUY-IN-DAY

- Reaffirmed Worth of QWL Teams
- Symbolized Management and Union's Coming Together in an Emotional Way
- Creative Energy at Its Best
- Fulfilled the "Impossible Dream"

CHAMPIONS

CHEMICAL QUAL-A-TEAM AND QWL TEAMS

Figure 8.4

At one point in the session, an East Indian team member, known affectionately to the rest of the team as "Babu," expressed the feeling that he shouldn't be there; everyone else looked shocked. Apparently, Babu felt that he hadn't done very much for the project. He hadn't made any speeches or organized any steering committees, and he generally felt that he didn't deserve this special recognition. But then the truth came out: Babu didn't speak

English very well. He explained to the team how clearly he had appreciated the fact that Buy-In-Day absolutely, positively had to succeed. He had felt afraid that his weak English might damage their chance of success, so he didn't get involved in project-specific tasks. In his words, "I just stayed back and worked."

What had really happened was that Babu had been doing everybody's job. While the team members were out making presentations and getting things organized, he'd been doing the work in the chemical purification department, making sure that no deadline was ever missed. Now everyone looked even more shocked and faintly embarrassed. Suddenly, the other team members rushed over to embrace him. All through the two months of preparation for Buy-In-Day, everyone had known that Babu was doing all their work in addition to his own, but no one had taken the time to say thank you. Nobody had ever acknowledged that he had made it possible for them to make Buy-In-Day happen. In the eyes of the team, Babu was the Champion that day.

Recognition in this form is a Celebration performed in the true spirit of the archetype. It occurs when people take the time to relive their journey. It allows them to express to each other the deep feelings that typically develop during a project's duration but rarely find release amid the chaos of getting the job done. This kind of Celebration allows people to express emotions that are simply too powerful and too touching to reveal in the workplace. It's simple, profound, and meaningful. It comes from the heart.

To conclude the session, both team members and facilitators expressed to one another their feelings about the session and their success story. The participants were asked to take a few minutes to reflect on the past six hours, and write down their feelings about the team renewal session.

One member of the team, James Porter, wrote:

> *I relived the feeling of the emotion at the bridging area. It reminded me of a TV program I saw once. An alien, unfamiliar with Christmas, gave his earth friends strange material presents. Upon seeing that the presents were inappropriate, he apologized. His friends reassured him, explaining that it wasn't the material gift that counted, but the thought behind*

it. With this new understanding of the 'thought,' the alien gave each of his friends a feeling, one that had been experienced at a very high point in each person's life. You had this power today, to make me relive this experience. Thank you for the present.

Marilyn and I were initially surprised by James' words. Six hours of painful stories felt like being given a present? But as we reflected on what James had written, we realized how powerful this kind of recognition can be and how very much we all need to feel valued.

With the victory of Buy-In-Day behind them, the team members returned to a normal work life. For some, it was a considerable letdown. Recognizing this, I challenged them to tackle a serious job-related project; it was one they'd been struggling with for months—a process to recover germanium. They accepted the challenge and within six months had a new process in operation, netting annual savings for the plant in excess of $250,000. This was followed by another letdown. So, the team came up with another big project, this time on its own; the team asked everyone in the plant to volunteer two hours a week for clean-up work, so we could turn the factory into a showplace. Both management and the union questioned the project and rejected the idea of asking people to work without pay. Undaunted, the Chemical Qual-A-Team went ahead with the project in its own operation and within months had transformed its work area into a showplace.

They found their next challenge completely outside their operation and began a water-treatment project that they hoped would impact the entire plant. This human cycle, taking them from one Impossible Dream to the next, is illustrated in Figure 8.5.

— ✧ —

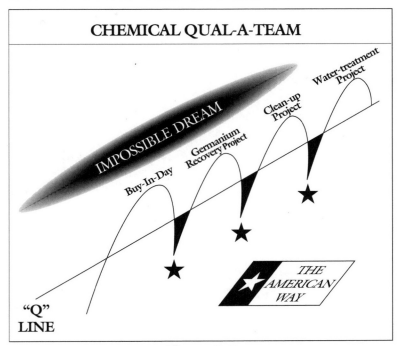

Figure 8.5

The members of the Chemical Qual-A-Team had learned the joy and power of working for an Impossible Dream, of being pulled forward by challenge. They had no intention of ever living without one.

I don't believe we can have justice without caring, or caring without justice. These are inseparable aspects of life and work, as essential for children as they are for adults.

—Judge Justine Wise Polier

— ✧ —

We must stop talking about the American dream, and start listening to the dreams of Americans.

—Reubin Askew

— ✧ —

Far away there in the sunshine are my highest aspirations. I may not reach them, but I can look up and see their beauty, believe in them, and try to follow where they lead.

—Louisa May Alcott

EPILOGUE

REDISCOVERING AMERICA

It's still true that in America, anything is possible—anything at all; but this is a concept that many Americans seem to have forgotten.

This country is full of men and women who never manage to reach their full potential; some dream dreams they never attain, while others never dream at all. Perhaps they need a little more support, a little more encouragement. Perhaps they need a special kind of nurturing, the kind that, in the past, has led Americans to excel to remarkable degrees—the kind that makes them dream with courage and work with devotion.

Everyone enters the adult world with emotional baggage that is outside their awareness. It can either sabotage your attempts at quality or can be made to work for your benefit. When you learn about the archetype, you learn about a part of yourself; it may be a part you never knew existed or at least never knew affected you to such an extent. Gaining such an insight into your self can be a moving, even profound experience. The deeper you probe and question the archetype, the more you'll recognize the depth to which it penetrates your life.

Our purpose on these pages has been to support the quest for a better, stronger America and a more rewarding American workplace.

We feel the ideas offered here can help rekindle that distinctive American spirit and promote the greater well-being of the individual and of society at large. It is our hope that this book will encourage people to express themselves: their feelings and their dreams. For everyone, the challenge is to help people maintain their enthusiasm, their sense of urgency, and their belief in the impossible.

The findings of the archetype study provide a solid foundation upon which to build an approach to quality and quality practices, one that is particularly suited to Americans. These findings are not limited to the specific problems faced by business, education, or government; rather, they suggest a timeless approach that can be applied to so many aspects of life throughout America. The archetype learning comes from our distinctive cultural background, and, when used with other established quality methods, it can provide an extremely powerful approach. The value of this knowledge is not in what you know, but in how you use it; it's not a matter of what you get, but how you give.

We believe there is a way to build a quality approach that will be valued by Americans because it is incredibly American, and its power comes from releasing the heart of quality. By aligning ourselves with our unconscious structure of quality, we Americans can help ensure this nation's preeminent position in the global marketplace. We must be free to be Americans—we must revel in our uniqueness and use the structures imprinted within us to achieve the results we all want.

Our work during the past five years has led us to believe that quality may be the unifying element that restores America's greatness and keeps alive the spirit that inhabits this land of opportunity. We invite you to pick a dream of your own. We ask you to stretch your imagination into the future, beyond presently accepted limits, to turn your dream into reality.

— ✧ —

Once there were two people looking out a window: one saw dirt ... the other saw stars.

JOIN OUR
QUALITY ARCHETYPE
SHARING NETWORK

Our purpose in writing this book has been to share our American quality archetype knowledge with people so that they might achieve more of their personal and organizational goals and, we hope, achieve some of their Impossible Dreams.

To that end, we desire to be a valued resource to you in your efforts to apply the learning from this study. We welcome your comments and questions, and we're eager to hear your experiences applying this new knowledge. We will continue to collect stories of practical applications and approaches that work. If you would like to join our network for sharing quality archetype experiences, please write to us at:

Hatala/Zuckerman
5721 Bend Creek Rd.
Dunwoody, GA 30338

BIBLIOGRAPHY

Ainsworth-Land, George T. *Grow or Die: The Unifying Principle of Transformation*. New York: John Wiley & Sons, 1986.

Bracey, Hyler, Jack Rosenblum, Aubrey Sanford, and Roy Trueblood. *Managing from the Heart*. New York: Delacorte Press, 1990.

Broholm, Richard R. *The Power and Purpose of Vision*. Indianapolis, IN: The Robert K. Greenleaf Center, 1990.

Bruner, Jerome S. *The Process of Education*. New York: Vintage Books, 1963.

Byham, William C. *Zapp! The Lightning of Empowerment*. Pittsburgh: Development Dimensions International, 1989.

Cramer, Kathryn. *Staying on Top When Your World Turns Upside Down*. New York: Viking Penguin, 1990.

Covey, Stephen R. *The Seven Habits of Highly Effective People*. New York: Simon & Schuster, 1989.

deBono, Edward. *Lateral Thinking: Creativity Step by Step*. New York: Harper & Row, 1970.

Deming, W. Edwards. *Out of the Crisis, 2nd ed*. Cambridge, MA: MIT Center for Advanced Engineering Study, 1986.

DePree, Max. *Leadership Is an Art*. New York: Bantam, Doubleday, Dell, 1989.

Exum, Ray. "My Favorite Football Story." *Reader's Digest*, September 1988, (Reprinted from the Chattanooga News—Free Press, October 27, 1982.).

Fallows, James. *More Like Us*. Boston: Houghton-Mifflin Co., 1989.

Fritz, Robert. *The Path of Least Resistance: Learning to Become the Creative Force in Your Life*. New York: Fawcett Columbine, 1989.

Gallwey, Timothy, and Bob Kriegel. *Inner Skiing*. New York: Random House, 1977.

Garfield, Charles. *Peak Performers*. New York: Avon Books, 1987.

Ginott, Haim G. *Between Parent and Child*. New York: Avon Books, 1956.

Guaspari, John. *The Customer Connection: Quality for the Rest of Us*. New York: Amacom, 1988.

Hamel, Gary, and C. K. Prahalad. "Corporate Imagination and Expeditionary Marketing." *Harvard Business Review*, July–August 1991.

Kamen, Mark. *The Karate Kid*, directed by John Avildson, Columbia Pictures Industry, 1984.

King, Martin Luther. "I Have A Dream." Speech delivered August 28, 1963, on steps of Lincoln Memorial in Washington, DC.

Kouzes, James M. *The Leadership Challenge*. San Francisco: Jossey-Bass Inc., 1987.

"Last Journey of a Genius," January 1989. (Produced for NOVA by WGBH TV, Boston.)

Ludeman, Kate. "Bosses Embrace Your Workers." *New York Times*, May 1989.

McGee-Cooper, Ann. *You Don't Have to Go Home Exhausted*. Dallas: Bowen & Rogers, 1990.

Mitroff, Ian I. *Business Not as Usual*. San Francisco: Jossey-Bass, 1987.

Moyers, Bill. "The Truth about Lies." The Public Mind, Public Affairs Television, Inc., November 29, 1989.

Naisbitt, John, and Patricia Aburdene. *Megatrends 2000*. New York: William Morrow & Co., Inc., 1990.

Neave, Henry R. *The Deming Dimension*. Knoxville, TN: SPC Press Inc., 1990.

Neill, A. S. *Summerhill.* New York: Hart Publishing Co., 1960.

Peters, Tom. *Thriving on Chaos.* New York: Alfred A. Knopf, 1987.

Peters, Tom, and Nancy Austin. *A Passion for Excellence.* New York: Random House, 1985.

Peters, Thomas J., and Robert H. Waterman, Jr. *In Search of Excellence.* New York: Warner Books, 1982.

Petersen, Donald E. From a speech delivered at the Conference Board's Marketing Conference, New York, November 1988.

Prange, Gordon W. *At Dawn We Slept: The Untold Story of Pearl Harbor.* New York: McGraw-Hill Book Co., 1981.

Rapaille, G. Clotaire. *The Creative Communication.* Paris: Dialogues, 1976.

Ryan, Kathleen, and Daniel K. Oestreich. *Driving Fear Out of the Work Place.* San Francisco: Jossey-Bass Inc., 1991.

Schaffer, Robert H. *The Breakthrough Strategy.* Cambridge, MA: Ballinger Publishing Co., 1988.

Senge, Peter M. *The Fifth Discipline.* New York: Doubleday, 1990.

Tichy, Noel M., and Mary Anne Devanna. *The Transformational Leader.* New York: John Wiley & Sons, 1986.

Whiteley, Richard C. *The Customer Driven Company: Moving From Talk to Action.* Reading, MA: Addison-Wesley Publishing Company, Inc., 1991.

Zais, Lt. General Melvin. From a speech delivered to the Army War College, Carlisle Barracks, PA, 1970.

ARCHETYPE BIBLIOGRAPHY
Prepared by Dr. G. Clotaire Rapaille

Benedict, Ruth. *The Chrysanthemum and the Sword,* Boston: Houghton-Mifflin Co., 1946.

Bettelheim, Bruno. *The Children of the Dream.* New York: MacMillan, 1969.

Bettelheim, Bruno, *On Learning to Read: The Children's Fascination with Meaning.* New York: Alfred A. Knopf, 1982.

Desmond, Adrian J. *Archetype and Ancestors.* Chicago: University of Chicago Press, 1984.

Evans, Richard I. and Konrad Lorenz. *The Man and His Ideas.* New York: Harcourt, Brace, Jovanovich, 1975.

Jung, Carl G. *The Archetype and the Collective Unconscious.* Princeton, NJ: Princeton University Press, 1986.

Jung, Carl G. *Four Archetypes.* London: ARK, 1972.

Lorenz, Konrad. *Behind the Mirror: A Search for a Natural History of Human Knowledge.* New York: Harcourt, Brace, Jovanovich, 1977.

Lorenz, Konrad. *Evolution and Modification of Behavior.* Chicago: University of Chicago Press, 1965.

Lorenz, Konrad. *Motivation of Human and Animal Behavior Imprinting: An Ethological View.* New York: Van Nostrand Reinhold Co., 1973.

McLuhan, Marshall. *From Cliche to Archetype.* New York: Viking Press, 1970.

Mead, Margaret. *New Lives for Old Cultural Transformations.* New York: William Morrow & Company, Inc. 1956.

Pearson, Carol S. *The Hero Within: Six Archetypes We Live By.* San Francisco: Harper & Row, 1986.

Piaget, Jean. *The Language and Thought of the Child.* New York: Meridian Books, 1955.

Piaget, Jean. *The Origin of Intelligence in Children.* New York: International Press, 1952.

Stevens, Dr. Anthony. *Archetypes: A Pioneering Investigation into the Biological Basis of Jung's Theory of Archetypes.* New York: Quill, 1983.

Ziegler, Alfred J. *Archetypal Medicine.* Dallas: Spring, 1983.

INDEX